MISSOURI GEOLOGY

T0141361

MISSOURI GEOLOGY

Three Billion Years of
Volcanoes, Seas, Sediments, and Erosion

A. G. UNKLESBAY
AND
JERRY D. VINEYARD

University of Missouri Press
Columbia and London

Library of Congress Cataloging-in-Publication Data

Unklesbay, A. G. (Athel Glyde), 1914–
    Missouri geology : three billion years of volcanoes, seas,
  sediments, and erosion / A. G. Unklesbay and Jerry D. Vineyard
        p.   cm.
    Includes bibliographical references and index.
    ISBN 978-0-8262-0836-1
    1. Geology—Missouri. I. Vineyard, Jerry D. II. Title.
  QE131.U54   1992
  557.78—dc20   92-7561
                                                         CIP

Designer: Rhonda Miller
Typesetter: Connell-Zeko Type & Graphics
Typefaces: Elante and Souvenir

*To George C. Swallow and the students
and science teachers of Missouri*

CONTENTS

PREFACE

Geology is the science that studies the earth. It builds on all the other sciences in the process of determining the present structure and composition of this planet and discovering its history through the more than three billion years of its existence. Geology is also the science charged with the responsibility of locating and finding ways to use the earth's rock and mineral resources for the welfare of mankind.

The geology of Missouri is fascinating and important, whether we look at it as scientists, economists, historians, sportsmen, or perhaps as artists. The answers to many questions regarding the geography, history, culture, and economy of Missouri can ultimately be found in the rocks and other geologic features of the state. Missouri provides a greater variety of geologic phenomena and geological resources than any of its neighbors.

The significance of geologic knowledge to human inhabitants began with the earliest prehistoric natives, who used the area's abundant flint for tools and weapons and who settled in communities along the rivers and around the springs and caves. They also knew about the salt springs and salt licks, which were important to them in food preservation and in treating animal skins. Throughout history, the rivers have provided convenient routes of travel; they were especially valuable to the early explorers, fur traders, and migrants seeking ways to the West.

Those who are familiar with the state often ask the same questions about the appearance of its surface. They want to know why there are so many caves in Missouri, and why all of the trout streams are in the south. They ask why the major cities and towns are located where they are, and why there are so many cement factories along the Mississippi and Missouri rivers in northeastern and southeastern Missouri and in the Kansas City area. Answers to these questions lie in an understanding of landscape features and the mineral content of the rocks beneath them.

Since the mid-nineteenth century, researchers have written many volumes on the geology of the state. Those works provide a wealth of details concerning the rocks, the minerals, the springs, the caves, the rivers, the hills, and the valleys. However, most of them were written by experts for other experts. This volume has borrowed much from them and is intended to be a synthesis of the fine work on Missouri in language that is understandable to the general reader. For anyone interested in learning more details about specific places or types of formations, the Bibliography at the end of this book provides selected references.

This book begins with a generalized description of the rocks of the state. It then provides descriptions of the state's physiography, caves, sinkholes, and water resources. We have attempted to use scientific terminology sparingly, but part of learning about the history of the earth is learning the words and phrases that describe it. A glossary is provided to help the reader through some of the more technical explanations. In addition, in order to discuss geology, it is essential to have access to a geologic time chart. Geologists use specific terms to refer to periods of time, beginning with the earliest part of earth's history, the Precambrian, and working through to the Quaternary, in which we live. We have provided charts to show general and specific time designations throughout the book, but it will be helpful to refer to the time chart in Figure 2–5 as you begin to read the introductory chapters. In addition, the generalized geologic map of the state in Plate 3 provides an overall view of the age of exposed features throughout the state.

The photographs in this book are intended primarily to show the larger features of rock formations. Because it is very difficult to study the distinguishing features of individual rock types through photographs, we have also provided a list of museums in the state where carefully labeled and identified specimens can be studied firsthand.

So, with photographs to show their general features and drawings to help explain how they came to be, we tell the story of Missouri's rocks and landforms. The story shows geology to be dynamic, ever-changing, foretelling future transformations through the history of the past.

ACKNOWLEDGMENTS

This volume builds on and is dedicated to the pioneering spirit of George C. Swallow, who came to the University of Missouri in 1851 to become chairman of the department of chemistry, geology, and mineralogy. According to his diary, it was a twenty-two-day journey from Brunswick, Maine, to Columbia by boat, railroad, and stagecoach, including three days to reach Columbia from St. Louis. In 1852 Swallow was named State Geologist, the first to hold this position. In 1872, he became the first Dean of the Agricultural and Mechanical College while also serving as professor of botany, comparative anatomy, and physiology in the medical school of the university. He was the author of the first five Geological Survey reports between 1855 and 1860. We also dedicate this volume to the science teachers of Missouri and to their students through whom continued research will be carried out for the benefit of the citizens of this state.

For assistance and support in the preparation of this volume, we gratefully acknowledge the help of many individuals and organizations. First, we must give credit to those whose geological research over many years provided information and ideas basic to the reading of Missouri's geologic history. Much of this was done by members of the Missouri Geological Survey, from its beginnings in 1853, to its present status as a division of the Department of Natural Resources. Of particular importance have been those whose help in more recent years has been useful: Arthur Hebrank, Wallace B. Howe, James A. Martin, Thomas L. Thompson, James E. Vandike, Heyward M. Wharton, John W. Whitfield, and James H. Williams. Members of the faculty at the University of Missouri who have helped are: Tom Freeman, Walter D. Keller, and James Stitt of the Columbia campus and Alfred Spreng of the Rolla campus. We must also recognize the classic works of E. B. Branson, whose comprehensive *Geology of Missouri* of 1918, revised in 1944, has long been a standard of reference.

We are also heavily indebted to the Missouri Department of Natural Resources (DNR), Division of Geology and Land Survey, for furnishing so many of the illustrations that help to explain and clarify geologic features and phenomena. Each of these is credited individually where it appears. Uncredited photographs and drawings were provided by A. G. Unklesbay.

The support of the University Press has been strong from the beginning of the project. Especially important has been the encouragement and support of Beverly Jarrett, Director and Editor-in-Chief; Jane Lago, Managing Editor, deserves special recognition for arranging the material in logical order and for making many technical phrases understandable. Susan Denny, former Associate Director, helped us over some rough spots in the very early stages. We also appreciate the help of Rhonda Miller, Designer.

We want to make it clear, however, that errors or incorrect statements are the responsibility of the authors.

MISSOURI GEOLOGY

1

ROCKS OF MISSOURI

The rocks of Missouri range widely in color, areas of exposure, mineral content, economic value, and commercial and scientific importance. They have attracted much attention from scientists, from mineral resource companies, from collectors, and from the general public. There are many rock and lapidary clubs scattered throughout the state. The Missouri General Assembly has recognized the importance of rocks by naming a state mineral (galena), a state rock (mozarkite), and a state fossil (a crinoid). Rocks have been the source of great mineral wealth and have provided the basis for quarrying and mining industries that have kept the state among the top ten nationally in mineral production. Currently, the annual value of the mineral industry of Missouri totals more than $4.5 billion.

Among the world's rocks, geologists recognize three major groups— igneous, sedimentary, and metamorphic. These groups are interrelated in the sense that, over time and given the right conditions, each type of rock can be transformed into one of the other types (Fig. 1–1). Missouri is almost exclusively composed of igneous and sedimentary rock types.

Igneous Rocks

The oldest rocks in Missouri are called *igneous* because they originated as a molten mass called *magma, ignis* being the Latin word for fire. From the evidence gathered by the geologists who have studied these rocks in detail, it now seems clear that the region of today's St. Francois Mountains was, nearly 2 billion years ago, a "hot spot." That is, it consisted of volcanoes emitting lava flows and spewing clouds of volcanic dust and ashes. Mixed with the ashes were quickly cooled fragments of magma that settled with the ashes of millions of fragments called *pyroclastics*. This mixture of magma and ash fragments cooled and solidified to form a rock known as *ash-flow tuff*. It is through several

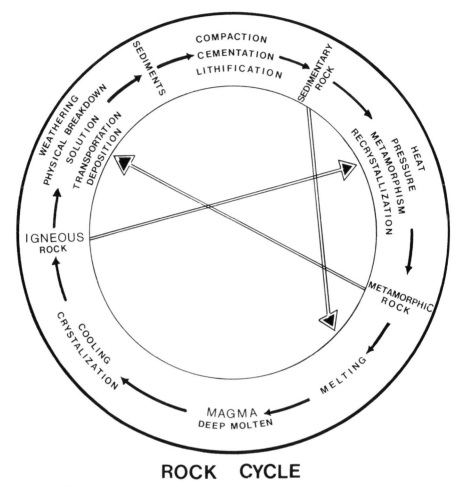

ROCK CYCLE

Fig. 1–1. The Rock Cycle represents the relationships among igneous, sedimentary, and metamorphic rocks. Natural processes are constantly transforming the rock. Weathering and transportation can be watched, but heating and compression take place deep within the crust.

layers of this resistant material that the East Fork of the Black River has cut the Johnson Shut-ins.

In addition to the tuffs the lava flows cooled quickly to form widespread masses of the glassy to fine-grained, hard, brittle, dark-colored rock known as *rhyolite*. One rhyolite in Iron County, named the Royal Gorge Rhyolite, has been dated as 1.53 billion years old. In some of the flows larger crystals grew before the mass completely cooled (Fig. 1–2). This created a dark, fine-grained mass with larger, light-colored crystals. This rock is called *rhyolite porphyry*. The larger crystals are called *phenocrysts*.

Fig. 1–2. A boulder of basalt porphyry showing large crystals within the rock's otherwise fine-grained matrix.

Other parts of the magma did not reach the land surface before cooling. Instead, the depth of overlying materials provided insulation, slowing down the cooling process enough for crystals to grow to sizes that can be seen without magnification. The resulting crystalline rock is called *granite*. Granite ranges in color from light gray to the more common red to pink. It is widely quarried and is the rock that forms the large, rounded, picturesque masses known as Elephant Rocks in the state park in Iron County (see Fig. 3–2). Because of regional differences in composition of the magma, different granites were formed. Three granites in Missouri have been dated. The Graniteville Granite in Iron County is 1.246 billion years old, and the Butler Hill Granite in St. Francois County and the Silver Mines Granite in Madison County are both 1.5 billion years old.[1]

Scattered through portions of the St. Francois Mountains is another variety of igneous rock called *basalt* or *diabase*. This rock is fine grained, dark gray to greenish black, and is very hard. It most commonly occurs in vertical tabular masses called *dikes* that cut through the granite and porphyry. These dikes formed as the molten magma invaded cracks or crevices in already existing igneous masses (Fig. 1–3). Structural relationships seem to indicate that the state's oldest igneous rocks are the rhyolites and tuffs, followed by the granites, and then the basalt masses.

All of these igneous rocks are quite resistant to weathering and erosion; thus they form the hills and knobs of the St. Francois Mountains. Especially prominent are Pilot Knob, Knob Lick, Iron Mountain, and Taum Sauk. The latter is the highest point in Missouri, 1,772 feet above sea level. Although these igneous rocks are known at the surface only in the St. Francois Mountains, they have been reached by drilling in many

Fig. 1–3. A diabase dike (the dark rock) cutting through older, light-colored granite in the St. Francois Mountain region of Missouri. DNR–Vineyard.

places and are believed to underlie the entire state. (For more about these rocks, see Chapter 6.)

Sedimentary Rocks

Sedimentary rocks are those composed of fragments derived by weathering and erosion of previously existing rocks or formed by chemical reactions and precipitation. The sedimentary rocks composed of fragments are referred to as "clastic," while the others are called "chemical." Clastic fragments are transported by wind or water and deposited as sediment in water or on land. The most common examples are the silt, sand, and mud carried by streams and deposited in lakes or oceans. Rocks can also be formed chemically by organisms such as corals building reefs, or by the accumulation of animal shells on the sea floor. Sedimentary rocks are usually classified by grain size. Clay and shale particles are those smaller than $1/16$ mm; sand-sized particles range from $1/16$ mm to 2 mm. Grains larger than 2 mm are called pebbles, cobbles, or boulders.

Missouri has a large assortment of sedimentary rocks that range in

Fig. 1-4. Road cuts for highways produce rock exposures that are extremely useful to geologists, who otherwise would not be able to see and study rocks covered by soil and other surface materials. Here a geology field trip stops to examine the massive cut at Cave Hill, along U.S. Highway 50 near Mount Sterling. Ordovician dolomites have been extensively deformed by groundwater solution, distorting the normally horizontal beds and forming sinkholes that were subsequently filled by clay. DNR–Vineyard.

age from Late Cambrian (see the geologic time chart, Fig. 2-5) to the present day, and from extremely fine-textured clays to coarse conglomerates. Beyond the St. Francois Mountains, the entire state is composed of layered, bedded, or stratified sedimentary rocks.

The most abundant of these are the marine carbonate rocks, limestone and dolomite (Fig. 1-4).[2] These are called carbonate rocks because they are composed of the minerals calcite (calcium carbonate) and dolomite (magnesium carbonate). Almost all of the Paleozoic-age rocks that cover Missouri south of the Missouri River belong to this category, and most are chemical or biochemical in origin. Those of Cambrian and Ordovician age (for example, what geologists call the Gasconade and Jefferson City formations) are dolomite; and the Devonian and Mississippian beds are limestone (one example is the Burlington Formation, illustrated in Plate 7). Geologists use the term *formation* to designate a rock unit that has uniform characteristics, is tabular in form, and is large enough to be mappable. Formations are commonly named for a geographic feature, or area, where they are first studied. The locality where a specific formation is first described usually becomes known as the *type section* or *type locality*.

Some of the dolomites in Missouri, such as Jefferson City Dolomite, are used for building stone. Bonneterre Dolomite has been the source

Fig. 1–5. This road cut along Highway 8 west of Flat River provides access to the Davis Shale and the overlying Derby–Doe Run Formation. DNR–Vineyard.

rock for most of the lead produced in the state. Most of the caves and springs of the Ozarks occur in dolomites of the Cambrian and Ordovician periods.

Limestones from the Mississippian Period are widespread from the northeastern to the southwestern parts of the state and are important commercially for building stone and agstone. Beds from the formations known as Burlington and Keokuk are quarried extensively. In the southwest region, those from the Warsaw Formation are cut and polished and sold on the market as Carthage marble. They are widely used for floors and walls in larger buildings and wherever polished ornamental stone is desired. In the Kansas City region, limestones of Pennsylvanian age form prominent cliffs in the city and are quarried for building stone, crushed stone, and stone used in making cement.

Also important to the state are several of the Ordovician and Cambrian sandstones. The earliest Cambrian formation is the massive fine- to coarse-grained Lamotte Sandstone, which is clearly derived from the underlying granite. In some places it contains enough coarse pebbles of the granitic material to be called a conglomerate. More recent are the medium- to fine-grained Gunter and Roubidoux sandstones, which are porous and serve as water sources to deep wells. Both are also used for building stones. Even more recent, but generally restricted to the east-central part of the state, is a third sandstone composed of exceptionally pure quartz with spherical grains and some cross-bedding, the result of shifts in the direction of the current as the sand was being deposited. This St. Peter Sandstone is quarried and mined in the Crys-

Fig. 1–6. Typical Missouri flint fireclay, one of the state's important mineral resources. DNR–Vineyard.

tal City and Pacific areas and is an important raw material for glass making.

Two other widespread sedimentary rocks in the state are shale and clay, which are often referred to as *mudstones*. Shale is a layered and laminated rock composed of a mixture of silt- and clay-sized particles (Fig. 1–5; see also Figs. 6–9 and 6–31). The beds may vary in thickness from a few inches to more than fifty feet. Most shale beds are light to dark gray, but some are yellow to red. Much of the shale that is mined is used in making brick and tile, but shale is also important as an ingredient in the manufacture of portland cement. The Maquoketa Shale of northeastern Missouri is especially important to the cement industry.

Clay differs from shale in that it is not layered and lacks the silt-sized particles. It is used for all sorts of ceramic products, especially brick and tile. Some varieties, known as *refractory clay* or *fire clay*, are used in making bricks or other blocks for lining blast furnaces and other structures requiring material with a very high melting point (Fig. 1–6). Missouri has long been a leader among the states producing this important material.

Ordinarily, when we think of sedimentary rocks, we think of sandstones, shales, and limestones. However, in Missouri we must also think of coal. Coal has been mined in fifty-five counties in Missouri, and it occurs over about 24,000 square miles in the northern and western parts of the state. More than forty different beds or *seams* have been identified, and they range in thickness from less than an inch to more than seventy-five inches. Of the forty beds, only about half are of minable thickness. All Missouri coal is Pennsylvanian in age and occurs in localities where swamp environments existed. The coal formed when plant remains became buried under conditions that led to the preservation of black, combustible carbon. Missouri's coal ranges in grade from place to place and from bed to bed but is all ranked as bituminous, which means that it contains organic or carbonaceous material and can be burned.

Within the rocks making up Missouri, there are many distinct or localized features. Many of these will be discussed and examined in greater detail in the following chapters.[3]

2

READING THE HISTORY

Before trying to "read the history from the rocks," we must recognize that the earth is constantly changing and that it has existed for a long, long time. We sing about the "Rock of Ages," and in most of our lifetimes we see only minor changes among the hills and valleys where we live. However, we do see the effects of floods, landslides, earthquakes, volcanoes, and hurricanes, evidence that the earth is an ever-changing planet.

Less obvious to us are the great changes over vast periods of time that cause major regions of a continent to rise and fall, to be submerged by the seas and then emerge again. Some of these changes are phenomenal in extent and can lift sea-floor sediments to the tops of mountain ranges or break entire continents into segments called *plates* and move them to a new position on the face of the planet. We term these movements *continental drift,* and the study of the plate movements is called *plate tectonics. Tectonics* is the study of the earth's broad structural features.

About 200 million years ago the continents were not separate masses as we now know them. They were all joined together in one large continental mass that we now call *Pangea* (Fig. 2-1). Ruptures appeared in several areas, and Pangea was split into two separate masses, each of which later split further into smaller ones. The northern of the two, called *Laurasia,* became the North American plate and Eurasia. The southern mass, referred to as *Gondwana,* became South America, Africa, and Australia. On a map of the earth you can see how well the east margins of North and South America fit the west margins of Africa and Europe. If you look at a map of the floor of the Atlantic Ocean, you will find a great rift or tear that marks the break-line between these continental masses. The current continents are still moving and separating, by current estimates at the rate of a little less than two centimeters (about three-quarters of an inch) in a year.

Over many millions of years, these movements of the continental

9

Fig. 2-1. Theoretical relations of the continents in Early Mesozoic time, about 200 million years ago.

masses and the individual plates have resulted in a warping of the earth's crust that has caused large areas of the continents to be periodically submerged and flooded and alternately raised and exposed to weathering and erosion. The collision of these plates causes volcanoes and earthquakes. These movements have allowed for the accumulation of a variety of materials and different kinds of sedimentary rocks. These layers of rock tell us the story of their formation. By learning the "language of the rocks," we can interpret the record of the earth's history.

The Midcontinent area in which Missouri lies has been flooded and drained many times. However, being in the middle of the continent has protected the area from the effects of collisions between the continental plates, making it relatively free of the resulting volcanic and earthquake activity. One exception to this general statement is the New Madrid seismic area, where there have been many earthquakes caused by deep-seated faults. These faults were probably formed in very early stages of the earth's history, when the New Madrid region was near the edge of a continental plate.

While it is true that most of the geologic history of the state is told in these massive plates of igneous rocks and the sedimentary rocks that lie upon them, the northern half of the state has a chapter of its history that is related not to crustal movement but to climatic change. During the last 2 million years, much of northern Europe and northern North America has undergone extreme changes in climate, resulting in at least four intervals when great continental ice sheets, or glaciers, formed in the far north and moved southward. The ice-forming intervals alternated with periods of warming when the ice melted away. Each time the ice moved slowly southward, like a great abrasive sheet, it gathered within itself tons of clay, sand, and gravel and even boulders, which were then left behind when the ice melted. Of the several great ice sheets, it seems only the early ones extended into the area that is now Missouri, reaching a position that almost coincides with the valley of the Missouri River, although a later one did extend a small tongue into the St. Louis region.

The Age of the Earth—How Old Is Old?

When early observers began to see that flowing streams carried sediment and realized that erosion was taking place, they began to wonder how long it had taken for various geologic features (hills and valleys) to be formed. This led to speculation about the span of geologic time. James Hutton, an early observer and writer in Edinburgh, wrote in 1788, "We find no sign of a beginning—no prospect of an end." This is probably the best statement ever to describe the immensity of geologic time. It is often said that one of geology's greatest contributions to the world of science is its demand for grasping the concept of vast amounts of time.

Techniques now exist to determine the ages of rocks that contain certain radioactive elements. To simplify, uranium over time will "decay" to form isotopes of lead. The rate at which this decay takes place has been accurately determined, providing a reliable way to estimate the age of any rocks containing these isotopes. The oldest rocks on earth for which these age determinations have been made are nearly 4 billion years old. Some meteorites have been determined to be 4.6 billion years old, so it seems safe to estimate that the earth is at least 4.5 billion years old. Given our perspective of human history, such a time span is difficult to grasp, but given the colossal changes made in the earth since its beginning, the age does not seem unreasonable.

Based on more than two thousand years of study and observation by many scientists, geologists have developed a chronological scale by which we can make time correlations between rock formations from around the earth. We can then interpret and understand the sequence of geologic events. It is customary to depict a rock sequence in a vertical *column*, with the older rocks underneath the younger ones. The *Law of Superposition* states simply that in a normal succession, it is logical to

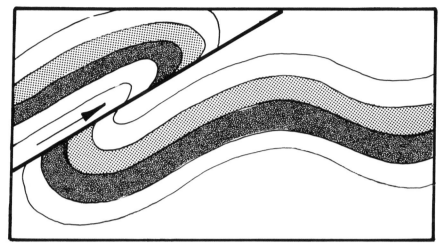

Fig. 2-2. Cross-section showing how faulting can cause the older rocks to be positioned above the younger ones.

assume that the lower layer was there when the upper was deposited over it. However, in areas where there has been extreme disturbance, entire rock sequences may have been overturned, reversing normal age relationships (Fig. 2-2), so the process of determining the relative ages of rocks is not quite as simple as it might seem.

As with most accounts of history, there is a need for a baseline or a plane of reference in talking about the comparative ages of rock formations. For example, we use the birth of Christ for our ordinary calendar, with historical dates designated as B.C. or A.D. Geologists divide time at the base of what is called the Cambrian sequence—that is, the lowermost rocks that contain well-preserved fossils of animals that possessed hard parts such as bones or shells. Accordingly, we deal with rocks that are Cambrian or younger, and those that are Precambrian. This boundary is recognized worldwide. The rocks at the base of the Cambrian are about 550 million years old (Fig. 2-3). Below the Cambrian throughout most of the world are igneous and metamorphic rocks of such complexity that it is not possible to accurately determine any details of their history. Some of the less deformed ones do bear evidence of soft-bodied forms of primitive life (Fig. 2-4). Although we still have much to learn about the beginnings of life on the earth, the record clearly indicates that life began long before the Cambrian Period.

Using this boundary as a reference point, geologists have established a "geologic calendar," which divides geologic time into periods (Fig. 2-5). These periods have been named for the geographic areas where

Fig. 2–3. This view of the Taum Sauk Power Plant shows layered Cambrian Lamotte Sandstone resting upon Precambrian igneous rock. DNR–Vineyard.

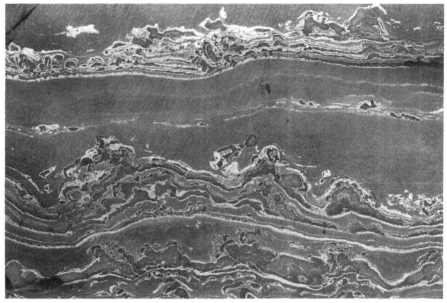

Fig. 2–4. Precambrian algal stromatolite sawed from a limestone in the St. Francois Mountain area. These are the oldest known fossils in Missouri. DNR–Bruce
Stinchcomb.

the rocks from each period were first studied, or for some characteristic
of the rocks. The names, from youngest to oldest, are listed and briefly
defined here:

Quaternary: This name is applied to the most recent 2 or 3 million
years since the end of the Tertiary, and includes the present time. Some
writers use *Holocene* for the most recent 10,000 years and *Pleistocene*
for the Ice Age—the time of extensive glaciation of Europe and northern North America.

Tertiary: This term is a carryover from the days when the basal or
lowermost igneous and metamorphic rocks were called *Primary,* the
early fossil-bearing sedimentary rocks were called *Secondary,* and those
rocks above the Secondary were called *Tertiary.*

Cretaceous: This name is derived from the Latin word *creta,* which
means chalk and was first applied to rocks in western Europe, where
there are extensive layers of chalky carbonate sediments.

Jurassic: These rocks were first extensively studied in the Jura Mountains of France and Switzerland.

Triassic: This is one of the few nongeographic terms that geologists
use. Literally, *triassic* describes a rock sequence clearly divided into three
distinct parts.

GEOLOGIC TIME CHART

TIME UNITS			YEARS AGO	CHARACTERISTIC LIFE
CENOZOIC ERA	QUATERNARY	PLEISTOCENE EPOCH	2,000,000	MODERN FORMS OF LIFE DEVELOPED
		PLIOCENE EPOCH	12,000,000	
		MIOCENE EPOCH	24,000,000	
	TERTIARY	OLIGOCENE EPOCH	37,000,000	
		EOCENE EPOCH	58,000,000	
		PALEOCENE EPOCH	66,000,000	
MESOZOIC ERA		CRETACEOUS PERIOD	160,000,000	APPEARANCE OF FLOWERING PLANTS / DINOSAURS COMMON
		JURASSIC PERIOD	206,000,000	MANY GANOID FISHES / FIRST BIRDS / DINOSAURS
		TRIASSIC PERIOD	245,000,000	FIRST MAMMALS / AMPHIBIANS, REPTILES, AND FISHES
PALEOZOIC ERA		PERMIAN PERIOD	285,000,000	REPTILES DIVERSIFY / AMPHIBIANS / INSECTS / MOLLUSCA
		PENNSYLVANIAN PERIOD	320,000,000	COAL PLANTS / FIRST REPTILES / FIRST INSECTS / MOLLUSCA
		MISSISSIPPIAN PERIOD	360,000,000	SHARKS / GREAT DEVELOPMENT OF CRINOIDS / COAL PLANTS
		DEVONIAN PERIOD	410,000,000	"AGE OF FISHES" / PRIMITIVE AMPHIBIANS / FIRST FORESTS / BRACHIOPODS
		SILURIAN PERIOD	437,000,000	FIRST CORAL REEFS / CRINOIDS ABUNDANT / FIRST SCORPIONS AND AIR-BREATHING VERTEBRATES
		ORDOVICIAN PERIOD	495,000,000	RISE OF CEPHALOPODS / FIRST PRIMITIVE FISH / CRINOIDS / GASTROPODS
		CAMBRIAN PERIOD	550,000,000	TRILOBITES / BRACHIOPODS / SPONGES
PRE-CAMBRIAN ERAS			3,350,000,000	INDICATIONS OF LOW FORMS OF ANIMALS AND PLANTS / ALGAE

Fig. 2–5. Generalized geologic time chart showing major subdivisions of time and typical representatives of the life of each subdivision.

Permian: So called because of extensive deposits in the province of Perm in Russia.

Carboniferous: This name was first used widely in Europe because these rocks contain many beds of coal and carbon-rich shales. The Carboniferous in the United States is commonly divided into an upper part called Pennsylvanian, for the thick section of coal-bearing rocks in that state, and a lower part called Mississippian, for exposures in the upper Mississippi Valley.

Devonian: These beds were first studied and described by British geologists and named for their extensive occurrence in Devonshire, England.

Silurian: This name is derived from an area in England once occupied by a Celtic tribe known as the Silures.

Ordovician: In a manner similar to the Silurian, this period derives its name from a region occupied by another tribe called the Ordovici.

Cambrian: This name was derived from Cambria, the Roman name for the area in the British Isles currently known as Wales.

Precambrian: This is the name given to any rocks dating from before the Cambrian Period. It is difficult to comprehend such vast amounts of time, but the oldest known rocks are about 4.5 billion years old. The span of time from the formation of these oldest rocks to the Cambrian is divided into two periods. The oldest extends from nearly 4 billion years ago to 2.5 billion and is called the *Archaean.* The rest of that time—until the Cambrian, from about 2.5 billion to 600 million years ago—is called the *Proterozoic.*

Many of the rock layers in the geologic column contain fossils that are distributed around the world. In such cases it is customary to make time correlations between widely separated regions. For example, certain fossils from the Pennsylvanian Period in Missouri can be matched with fossils from the Upper Carboniferous Period of England. Permian fossils from Texas can be matched with similar forms in Russia and in South America.

Geologists have also grouped these period names into larger time units, called *eras.* This grouping is based on general features of the fossils that represent the characteristic life of each time. The names of the eras can be defined as follows:

Cenozoic: recent life.
Mesozoic: middle life.
Paleozoic: ancient life.
Proterozoic: primitive life.
Archaean: before the dawn of life.

These terms will help us to understand the rocks and surface features of Missouri, which we will examine in chapters that follow.

A first consideration of the enormous amount of time represented by these named time units causes one to be almost overwhelmed. To think of the continents of South America and Africa drifting apart by the width of the Atlantic Ocean in a period of some 200 million years calls upon us to stretch our imaginations. However, when we examine the "record in the rocks" and find that fossils that represent sea-floor life are now high above sea level in Missouri and that rocks of similar age and origin are even higher in the Rocky Mountains, we must accept the fact that great earth movements have occurred that did not happen quickly.

3

THE SURFACE OF MISSOURI

If we could look down on Missouri from a spaceship, we would see a variety of landforms. As a substitute for this view from space, this book contains both shaded and colored relief maps, which graphically show the different forms of terrain. If the relief maps (Plates 1 and 2) are compared with the geologic map (Plate 3), it becomes obvious that there is a definite relationship between the landforms and the underlying rock types; the landforms are the result of the geologic history of the region.

In our view from the sky, the most distinctive feature of the state would be the broad, domelike shape of southeastern Missouri (Fig. 3–1). The Precambrian rocks of the St. Francois Mountains form the highest area in the state, with Taum Sauk Mountain in Iron County, at 1,772 feet above sea level, being the highest point. Dipping away from the dome created by the mountains are Paleozoic sedimentary rocks. The U.S. Geological Survey gives the approximate mean elevation of Missouri, or the average of all the elevations in the state, as 800 feet above sea level. The lowest elevation is 230 feet, in Dunklin County along the St. Francis River. The geographic center of the state, or the point at which the state would balance if it were a flat plane, is in Miller County, about twenty miles southwest of Jefferson City.

Plate 1 shows five major regions, or provinces, of surface landform development in Missouri: the St. Francois Mountains, the Ozarks, the Southeastern Lowlands, the Western Plains, and the Glaciated Plains. As used by geologists, the term *province* refers to an area that is characterized by particular geologic or geomorphic features. The features that distinguish each of the five provinces in Missouri will be discussed in turn.

The St. Francois Mountains

This rugged, rocky part of the state contains the oldest exposed rocks in the Midcontinent region. These igneous, volcanic rocks are Precambrian

18

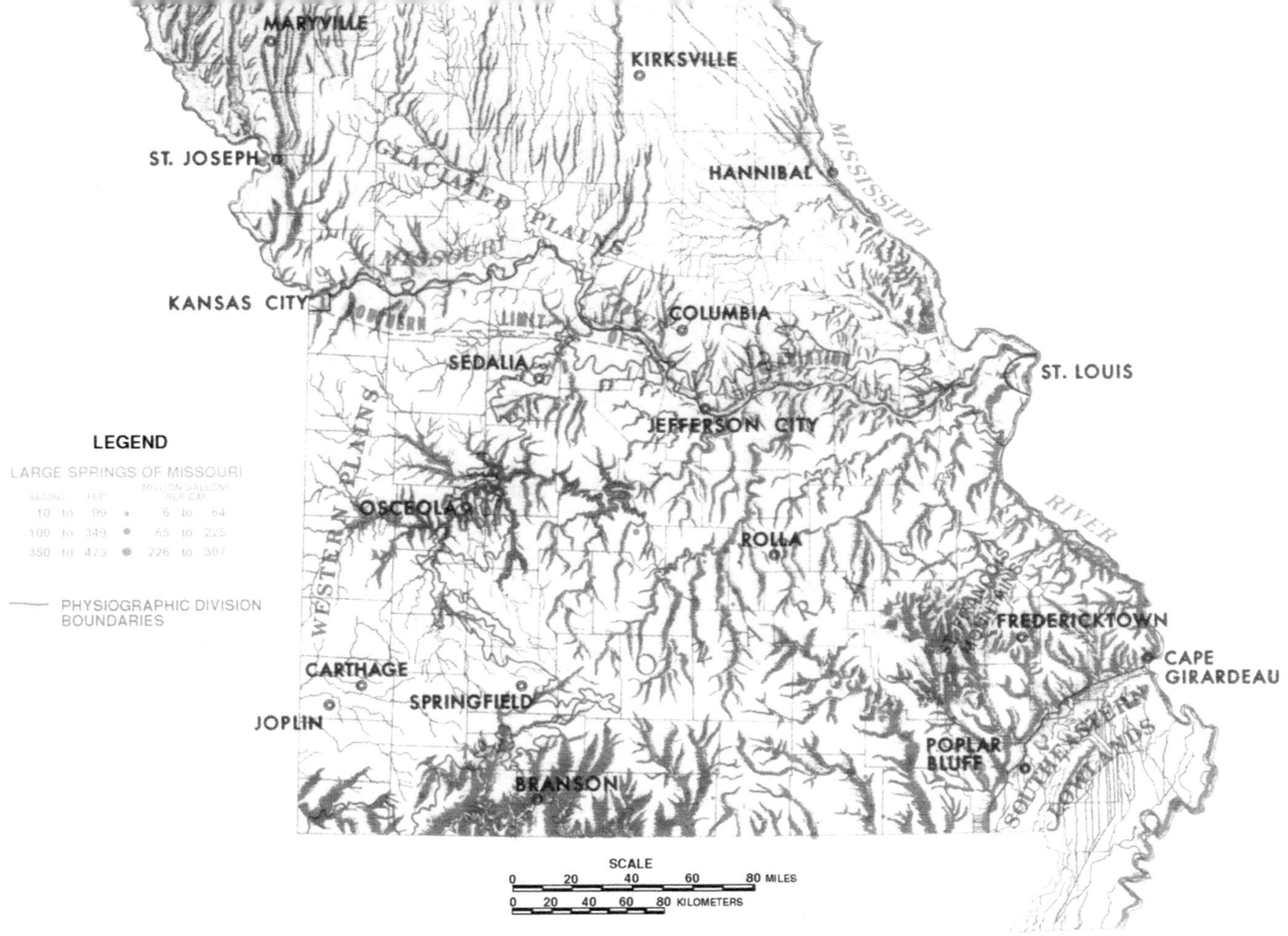

Plate 1. Relief map of Missouri. DNR.

Plate 2. Land relief of Missouri. DNR.

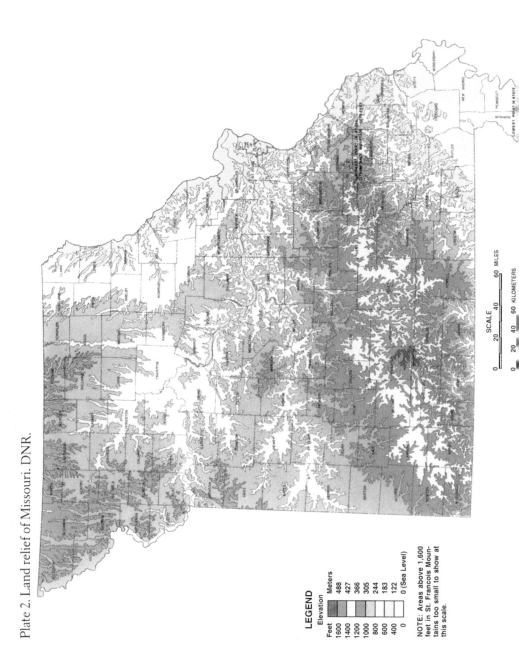

LEGEND

Elevation	
Feet	Meters
1600	488
1400	427
1200	366
1000	305
800	244
600	183
400	122
0	0 (Sea Level)

NOTE: Areas above 1,600 feet in St. Francois Mountains too small to show at this scale.

SCALE

0   20   40   60 MILES

0   20   40   60 KILOMETERS

20

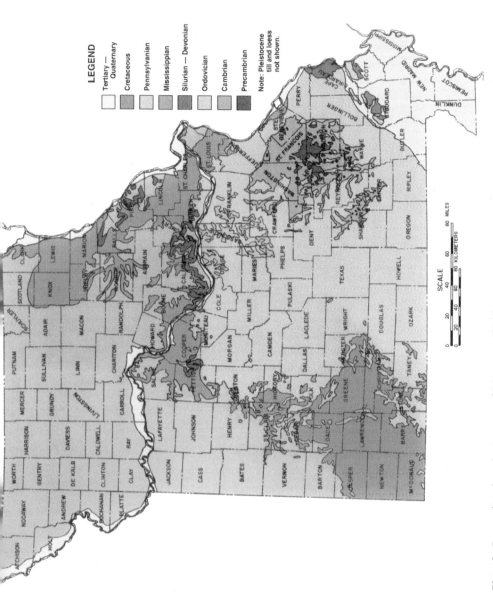

Plate 3. Generalized geologic map of Missouri. DNR.

Plate 4. Surface materials map of Missouri. DNR.

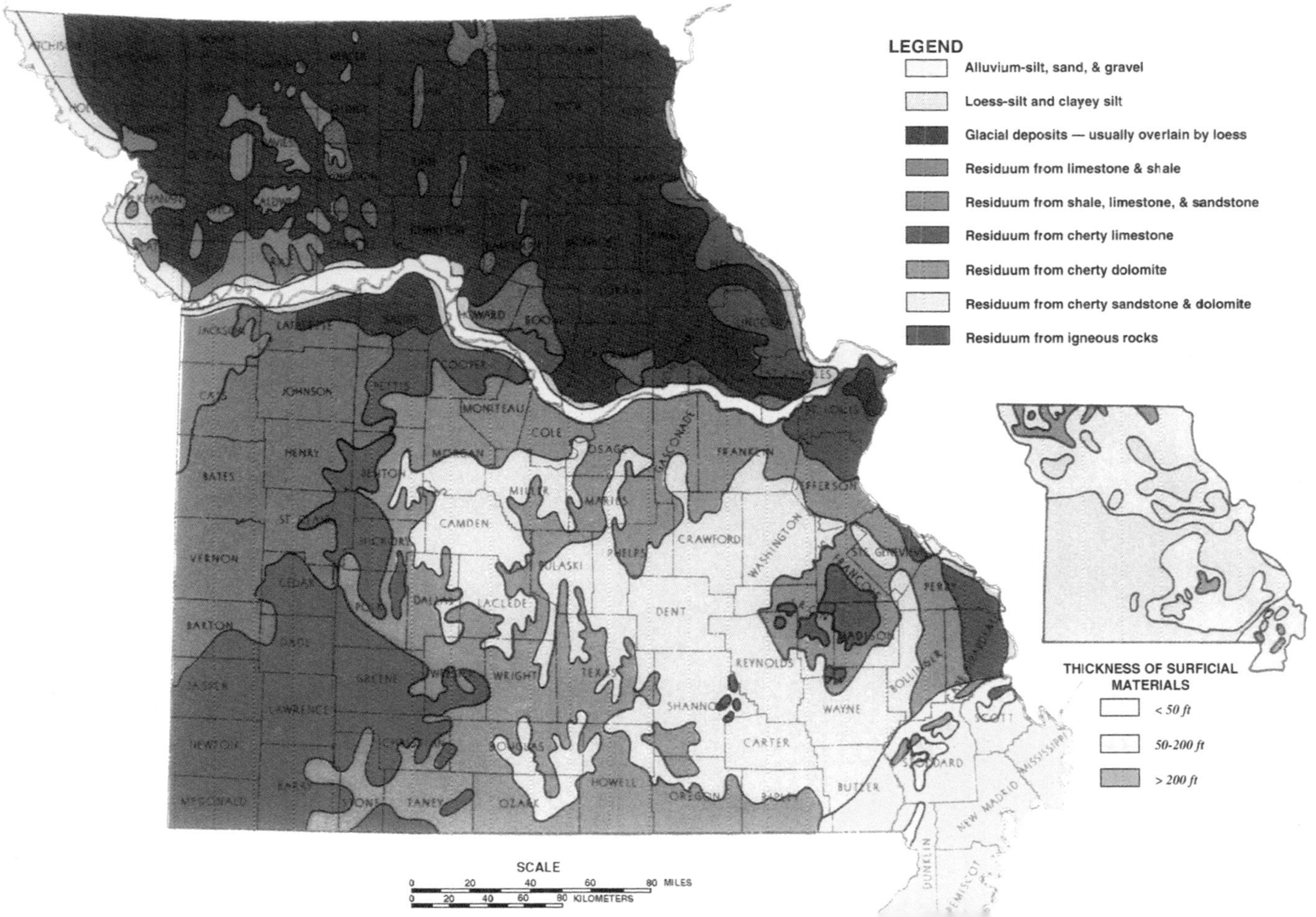
LEGEND
Alluvium-silt, sand, & gravel
Loess-silt and clayey silt
Glacial deposits — usually overlain by loess
Residuum from limestone & shale
Residuum from shale, limestone, & sandstone
Residuum from cherty limestone
Residuum from cherty dolomite
Residuum from cherty sandstone & dolomite
Residuum from igneous rocks
THICKNESS OF SURFICIAL MATERIALS
< 50 ft
50-200 ft
> 200 ft
SCALE
0
20
40
60
80 MILES
0
20
40
60
80 KILOMETERS
JACKSON
JOHNSON
HENRY
BATES
VERNON
BARTON
JASPER
NEWTON
MCDONALD
HOWARD
COOPER
MONITEAU
COLE
MORGAN
CAMDEN
MILLER
OSAGE
MARIES
GASCONADE
FRANKLIN
JEFFERSON
CRAWFORD
WASHINGTON
PHELPS
PULASKI
LACLEDE
DENT
REYNOLDS
SHANNON
CARTER
WAYNE
BUTLER
OREGON
HOWELL
OZARK
TANEY
WRIGHT
GREENE
DOUGLAS
PERRY
BOLLINGER
SCOTT
NEW MADRID
MISSISSIPPI
DUNKLIN
PEMISCOT

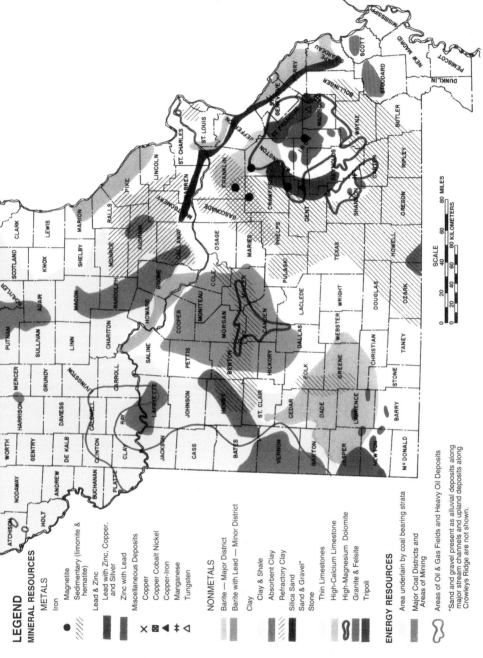

LEGEND

MINERAL RESOURCES

METALS

Iron
- Magnetite
- Sedimentary (limonite & hematite)

Lead & Zinc
- Lead with Zinc, Copper, and Silver
- Zinc with Lead

Miscellaneous Deposits
- ✕ Copper
- ⊠ Copper Cobalt Nickel
- ◀ Copper-Iron
- ‡ Manganese
- ◁ Tungsten

NONMETALS

- Barite — Major District
- Barite with Lead — Minor District

Clay
- Clay & Shale
- Absorbent Clay
- Refractory Clay

- Silica Sand
- Sand & Gravel*

Stone
- Thin Limestones
- High-Calcium Limestone
- High-Magnesium Dolomite
- Granite & Felsite
- Tripoli

ENERGY RESOURCES

- Area underlain by coal bearing strata
- Major Coal Districts and Areas of Mining
- Areas of Oil & Gas Fields and Heavy Oil Deposits

*Sand and gravel present as alluvial deposits along major stream channels and upland deposits along Crowleys Ridge are not shown.

SCALE

0   20   40   60   80 MILES

0   20   40   60   80 KILOMETERS

Plate 5. Mineral and energy resources in Missouri. DNR.

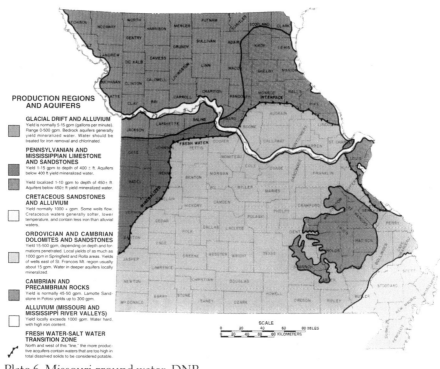

PRODUCTION REGIONS
AND AQUIFERS

GLACIAL DRIFT AND ALLUVIUM
Yield is normally 5-15 (gallons per minute).
Range 0-500 gpm. Bedrock aquifers generally
yield mineralized water. Water should be
treated for iron removal and chlorinated.

PENNSYLVANIAN AND
MISSISSIPPIAN LIMESTONE
AND SANDSTONES
Yield 1-15 gpm to depth of 400 ± ft. Aquifers
below 400 ft yield mineralized water.

Yield localized 1-10 gpm to depth of 450± ft.
Aquifers below 450± ft yield mineralized water.

CRETACEOUS SANDSTONES
AND ALLUVIUM
Yield normally 1000 + gpm. Some wells flow.
Cretaceous waters generally softer, lower
temperature, and contain less iron than alluvial
waters.

ORDOVICIAN AND CAMBRIAN
DOLOMITES AND SANDSTONES
Yield 15-500 gpm, depending on depth and formations penetrated. Local yields of as much as
1000 gpm in Springfield and Rolla areas. Yields
of wells east of St. Francois Mt. region usually
about 15 gpm. Water in deeper aquifers locally
mineralized.

CAMBRIAN AND
PRECAMBRIAN ROCKS
Yield is normally 45-50 gpm. Lamotte Sand-
stone in Potosi yields up to 300 gpm.

ALLUVIUM (MISSOURI AND
MISSISSIPPI RIVER VALLEYS)
Yield locally exceeds 1000 gpm. Water hard,
with high iron content.

FRESH WATER-SALT WATER
TRANSITION ZONE
North and west of this "line," the more productive aquifers contain waters that are too high in
total dissolved solids to be considered potable.

Plate 6. Missouri ground water. DNR.

Plate 7. Burlington Limestone showing bedded and nodular chert along Stadium Boulevard in Columbia.

24

Plate 8. A boardwalk leads visitors through the natural bridge at Rock Bridge Memorial State Park in Boone County. DNR–Nick Decker.

Plate 9. The Angel's Shower in Ozark Caverns is unique. DNR–Nick Decker.

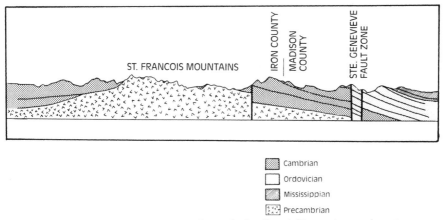

Cambrian
Ordovician
Mississippian
Precambrian

Fig. 3-1. An east-west cross-section through the Ozark Dome in southeastern Missouri.

in age (2 to 3 billion years old) and are shown on the geologic map in red. Although they are now at the surface, their texture and mineral content tell us that most of these rocks were originally emplaced as molten masses of magma that cooled and crystallized while deeply buried beneath the surface. Other parts of the igneous rock formations are glassy and apparently reached the surface as volcanic ash and lava flows. We do not know what kind of material preceded igneous activity, because it was removed by erosion over a period of perhaps a billion years, during which a rough topographic surface was developed. This surface had many deep valleys and prominent knobs, some of which still exist in the St. Francois Mountain area and are known elsewhere in the state from deep drilling.

The rocky knobs of the area are composed mostly of red and gray granite, but other igneous and volcanic rocks are also present. Two other fairly common rocks are called *rhyolite* and *felsite*. They have a dense to glassy texture and apparently cooled nearer to the surface. Some parts may have actually reached the surface as lava flows. In some places the granites have been invaded by later intrusions of the dark basaltic rock called *diabase*, which occurs as dikes cutting through the granites (see Fig. 1-3). Among the knobs of igneous rock are Pilot Knob, Iron Mountain, Knob Lick, and Taum Sauk. These Precambrian rocks have been important as iron ores and as a possible source of lead, zinc, copper, and other rare elements and precious metals that were carried into the Cambrian rocks in aqueous solutions during a post-Cambrian upward movement of the St. Francois Mountains.

About 520 million years ago the earth's crust warped, and seas began

Fig. 3-2. Elephant-sized granite boulders in Elephant Rocks State Park near Graniteville. DNR–Vineyard.

to invade the region and may have completely covered the state. In fact, our reading of the rocks tells us that the region has been alternately submerged and exposed many times.

When the seas first invaded in late Cambrian time, the granite knobs were eroded, and the resulting sand and other rock fragments were deposited by the water to become what geologists call Lamotte Sandstone. As the seas deepened, carbonate sediments, especially limestone, began to accumulate on the sea floor. Successive layers of sediment spread widely over the region as a slow and constant overall rising of the sea level continued. Some of the knobs may have remained as unsubmerged islands; these appear now to have been exposed for millions of years. Keep in mind that during all these processes, land-plant life had not yet come into being.

Elephant Rocks State Park in Iron County is named for a large collection of granite boulders that has been exposed for a very long time, even in geologic terms (Fig. 3-2). Widespread masses of sediments still exist in the region, and upon them the rugged topography of the Ozarks has developed.

The Ozarks

The term *Ozarks* is used nationwide to refer to the hilly country of southern Missouri and northern Arkansas. As used here, the term designates the area south of the Missouri River, west of the St. Francois

Mountains, east of the Western Plains, west of the Mississippi River, and northwest of the Southeastern Lowlands. The western boundary is customarily drawn at the edge of the outcrop pattern of the Mississippian rocks that can be seen in Plate 3.

As Plate 3 shows, the Ozarks region consists primarily of Cambrian and Ordovician rocks, with limited areas of Pennsylvanian. The rocks of the area are for the most part "flat lying" to nearly horizontal but do dip slightly outward from the St. Francois Mountains. The repeated rising and falling of the sea created a series of repetitive sandstones, limestones, and dolomites, many of which contain fossil evidence of their marine origin. The total thickness of the sedimentary rocks is about 4,000 feet, and dolomites predominate.

The uplift and erosion of the last 300 million years have resulted in a topography of hills, plateaus, and deep intervening valleys. The valleys cut down through the rocks and expose them in many cliffs. As geologists study these exposures and those from the more recent road cuts and plot them on maps they find that they represent a gentle, widespread dome structure. Many of the hills range up to 1,000 to 1,200 feet above sea level. Here also are the deepest valleys in the state, many of which contain rivers or streams that have been dammed to create lakes. The best known of these are the Lake of the Ozarks, Harry S. Truman Reservoir, Stockton Lake, Pomme de Terre Lake, Lake Taneycomo, and Table Rock Lake. The old meander patterns of the streams give the lakes similar patterns with very long shorelines.

In the Ozark province, layered sedimentary beds have become the host rocks for extensive caves, springs, and sinkholes. Also throughout the Ozarks, the dolomites and limestones contain large amounts of chert (flint). The chert is not readily soluble; in areas where the dolomite and limestone have been dissolved and removed by weathering, the residual chert remains scattered over the surface. In some hilly areas erosion has removed the soluble rock leaving only a thin cherty soil.

Because there have been periods of crustal movement, the rocks in some localities have been folded and "wrinkled," as well as broken, or faulted. A major uplift at the close of the Ordovician Period brought most of the area out of the sea and exposed it to erosion during most of the 30 million years of the Silurian Period, and in addition much of the Devonian. Hence the region exhibits what geologists call an unconformity between the Ordovician and Middle Devonian beds (Figs. 3–3 and 3–4). That is, a significant period of time is not represented in the geologic layers, implying a substantial length of time during which there was no deposition of sediments. Apparently the area now occupied by

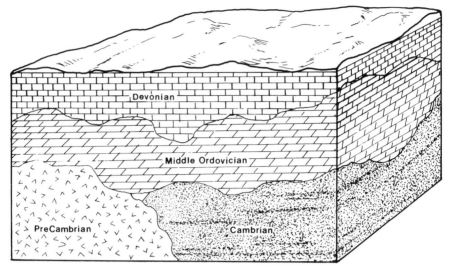

Fig. 3–3. Geologists use the term *unconformity* to designate a significant age gap between two rock layers. It usually implies a considerable length of time during which there was no deposition, or great erosion. In this case, the uneven surfaces between these rock masses represent great gaps in time.

Osage, Franklin, Gasconade, Maries, Crawford, and Phelps counties was again submerged during Pennsylvanian time, because Pennsylvanian sediments rest directly on Ordovician beds, as can be seen on the geologic map in Plate 3.

The Southeastern Lowlands

Markedly different from the rest of the state is the province known as the Southeastern Lowlands. This region is also sometimes referred to as the *Mississippi Embayment*. It contains the lowest point in Missouri, where the St. Francis River leaves the state at an elevation of 230 feet above sea level in Dunklin County. On the geologic map (Plate 3), this area is colored yellow. The same color is also used for the more recent alluvial, or flood, plains of the major rivers.

The relatively flat lowland area is composed mostly of river sediments that accumulated in a deep, subsiding basin, which is really an old channel of the Mississippi River. The basin has a long history of intermittent periods of subsidence. It began as a subsiding basin during Paleozoic time and continued to receive sediments into Cretaceous and Tertiary time. During these millions of years, the subsiding area was a northward-

4. CHOUTEAU: Early Mississippian

3. CALLAWAY: Middle Devonian

2. ST. PETER: Middle Ordovician

1. JEFFERSON CITY: Early Ordovician

Fig. 3–4. Exposure of three major unconformities along U.S. Highway 63 in southern Boone County. The bold lines represent great time gaps.

31

extending arm of the Gulf of Mexico. If Cape Girardeau had existed then, it could have been a seaport.

After the close of Tertiary time, the Gulf of Mexico retreated and the ancestral Mississippi and Ohio rivers eroded and removed much of the accumulated sediment, leaving remnants that Thomas Beveridge (1978) has called "lost hills." The largest of these is Crowley's Ridge in Stoddard and Scott counties. During the Ice Age (which geologists call the Pleistocene Epoch, see Fig. 2–5), the rivers, swollen by meltwater, again nearly filled the Embayment but did not cover the "lost hills." It is in them that we find the depositional record of the Cretaceous and Tertiary seas. The Embayment is still an unstable area and is in the center of the New Madrid earthquake zone. The relation between the subsidence and the occurrence of the earthquakes is not clear.

The Western Plains

The name *Western Plains* is applied to the area of Missouri south of the Missouri River and northwest of Springfield. It is the area shown on the geologic map (Plate 3) as being underlain by Mississippian and Pennsylvanian rocks. The Mississippian beds are mostly limestone, while the Pennsylvanian rock contains alternating beds of limestone, shale, and coal, with interlayered petroleum-bearing sandstones. The layered rocks of the area dip gently westward as part of the great domal structure of the state. This region has only minor structural features.

The area of Mississippian rocks is sometimes referred to as the Springfield Plateau. It is primarily an area of low relief—that is, without high hills or deep valleys but with broad, relatively level prairies. Warm seas with abundant marine life once covered the plateau, creating an abundance of thick, fossil-bearing limestones (Figs. 3–5 and 3–6).

The area underlain by Pennsylvanian rocks is sometimes called the Osage Plains. Its topography consists of low, rolling plains with broad, shallow valleys. The succession of limestones and shales with interlayered coal beds indicates a frequent warping of the crust and thus many shallow marine invasions alternating with terrestrial coal-forming swamps. Similar sedimentary records would develop if swampy areas like the Big Cypress swamp in Florida, and the Okefenokee Swamp in Georgia were to be alternately invaded by the sea and drained.

Fig. 3–5. A close view of the weathered surface of a crinoidal bed in Burlington Limestone. The small rounded discs are fragments of crinoid stems. They are about the size of a man's shirt button.

Fig. 3–6. Sketch of a modern crinoid. Much of Missouri's limestone is made up of fragments (columnals) of the stem.

The Glaciated Plains

Almost all of Missouri north of the Missouri River was covered at least twice by continental ice sheets. The Missouri River marks the approximate edge of the area that was covered by the ice, and meltwater flowing from the glaciers probably carved the original valley for the river. At one stage, the ice did temporarily extend a little farther south into parts of what are now Saline, Cooper, and Moniteau counties.

Before the ice moved southward into the area, the land surface had been subjected to erosion, and some valleys as deep as 300 feet had been formed. When the ice sheets receded, they left behind widespread deposits of clay, sand, gravel, and boulders (called *till* and *drift*). This is material that was scraped away from Iowa, Minnesota, and southern Canada. A more or less featureless plain was left by the glacial melting. In the 20,000 years since the last ice melted, new streams have formed new valleys unrelated to the older ones. West of a north-south line from northern Boone County to the Iowa border, the streams drain mostly southward to the Missouri River, and east of the line and north of Callaway County the drainage is mostly eastward to the Mississippi. These widespread layers of glacially deposited material provided the Glaciated Plains with the excellent farmland soils of today.

In Ralls, Pike, and Lincoln counties, and in southern Boone, Callaway, Montgomery, and Warren counties, the glacial drift has been removed by post–Ice Age erosion, and Paleozoic rocks are exposed.

There is a popular song that says "Everything is beautiful in its own way." Fortunately we all have different tastes in what we think beautiful, but when it comes to landscape beauty, Missouri, thanks to its geologic history, provides a great variety. In fact it provides a greater range in topographic development and a broader range in the geologic age of its rocks than any of its neighboring states.

It has been asserted by many historians and other authors that early pioneers and explorers were attracted by the state's natural beauty. Even now this beauty attracts thousands of visitors. A recent publication by Paul Nelson (1987) describes in detail many of the terrestrial natural communities of the state. Whether one likes forests or prairies, rocky glades or wetlands, clear spring-fed trout streams or sandy lake beaches, Missouri has them. It is an often overlooked fact that all these attractive features owe their origin to the state's geologic history.

4

STRUCTURAL FEATURES

To the geologist the widespread layers of rock that underlie the visible landforms are pages in the history of the state. Each layer has a story to tell. Its texture, its bedding, its mineral content, and its fossils document its origin and its history. It is interesting to think of these layers, or beds, as a stack of widely spread blankets. In some places they are worn thin, in others they are rumpled or folded, and in others they are filled with holes. In some places there are ragged edges (Fig. 4-1).

Although the Midcontinent region of which Missouri is a part is relatively stable, localized areas have undergone several periods of uplift and distortion since Precambrian time. The state's geologic features have been and continue to be affected by those uplifts. In order to discuss the specific structural features of Missouri, we need to define some terms and concepts and to remember that rock formations, even very thick ones, can be deformed by folding and fracturing. The earth's crust has been warped in many ways many times in response to being squeezed and stretched, raised and lowered, and individual rock beds, and sets of beds, have been bent, folded, broken, or otherwise deformed.

One of the common geologic structures is the *dome*, which can be likened to an upside-down bowl or shallow dish. Such structures are more or less symmetrical and roughly circular in overall shape. Some may be small, say a mile or less across, but others can be as large as a state. The Ozark dome in southern Missouri is an example of the latter (see Fig. 3-1).

Another feature is an *anticline*, in which one or more beds are folded convexly upward, but in an elongated rather than a circular pattern. Thus, the beds dip outward from a central axis. Anticlines may be small in scale or large enough to cover several counties or large regions of a continent. The converse to an anticline is called a *syncline*. In a syncline, the fold is downward, with the beds dipping in toward the axis. Synclines too are usually elongate in pattern. Anticlines and synclines

Fig. 4–1. Variation in bedding of the Jefferson City Dolomite on Highway 50, west of Jefferson City.

commonly occur together in areas where the crust has been broadly compressed (Fig. 4–2). A good example is the folded Appalachian Mountains.

When the rocks of an area being deformed are brittle, they may break instead of fold, and the result is a *fault* (Fig. 4–3). If the area is being stretched, one side of the fault will slide down along the fault plane. This is called a *normal fault*. If the area is being squeezed, the opposite will happen, and one side of the fault will be thrust upward. This is called a *reverse fault* or *thrust fault* (Fig. 4–4). Both kinds of faults may occur on a small scale or in some cases may be on a continental scale.

How Structures Are Identified

Features such as anticlines and synclines were first noted in field observations of rock beds dipping at various angles from the horizontal. By plotting the locations of exposed beds on maps and measuring the angle and direction of the dip in many places, geologists could draw lines to connect numbers and thus map the folds of the individual beds. Anticlines and synclines can also be determined by noting the relative ages of the rock layers. In anticlines that have been eroded, the central area will be composed of rock that is older than the surrounding beds, and in the synclines the opposite will be true (Fig. 4–5).

Underground structures can be detected and mapped in a similar way by plotting information gained from drilling for samples through a

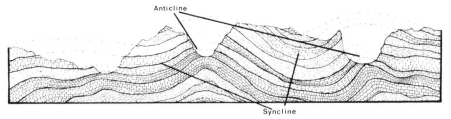

Fig. 4–2. Theoretical cross-section through an area of folded rocks to show anticlines and synclines.

Fig. 4–3. Faulting in the Eureka–House Springs structure along Highway 30 in Jefferson County. DNR–Vineyard.

sequence of rock beds. Supposed structures too deeply buried to allow either of these methods can be detected and mapped using seismic techniques. To do this, geologists lay out a pattern of instruments that will detect vibration in the rocks. Then, from a properly selected location, a shock wave is created that will penetrate the rocks of the area and be reflected back to the detectors. This shock wave may be created by a dynamite blast in a drill hole or by thumping the surface using a heavy weight. The shock wave will pass through different rock beds at different speeds and will send characteristic responses to the detecting instruments, which will plot and record them on graphs. From a study of the graphs the geologist can interpret the underlying structure. A common name for this procedure is *seismic reflection*. It is really a method that uses artificially created "earthquake" waves.

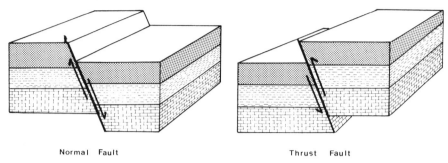

Normal Fault   Thrust Fault

Fig. 4–4. Block diagram showing the difference between a normal fault and a thrust, or reverse, fault.

PENNSYLVANIAN
MISSISSIPPIAN
DEVONIAN
SILURIAN
ORDOVICIAN
CAMBRIAN

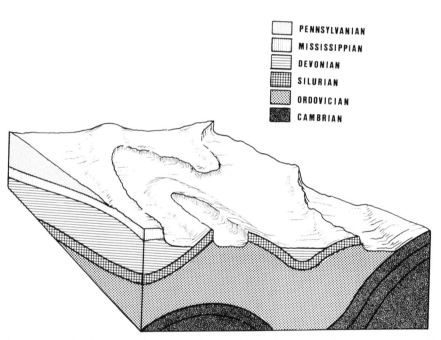

Fig. 4–5. Erosion in the folded areas of an anticline shows the older layers exposed near the center, while in a syncline the younger beds are exposed near the center.

*   *   *

Now we can return to the geologic map of Missouri (Plate 3) and look at some of the most prominent structures. As we noted in Chapter 3, in a general sense Missouri forms a broad dome, on the flanks of which are many smaller structures. The center of the dome is the St. Francois Mountains in the southeastern quarter of the state. The rocks there have been described earlier as a complex mass of igneous and metamorphic rocks of Precambrian age. Outward from this center, the layered sedimentary rocks of the Paleozoic dip away in all directions like so many blankets and sheets covering a sleeper. Except for the southeastern portion, the dips are gentle and hardly visible to the ordinary observer. To the southeast, the dips are steeper as the beds plunge into the subsiding area of the Southeastern Lowlands.

To the west and southwest the beds dip gently across the state and disappear under Kansas and northern Arkansas. To the northeast, they spread out under Iowa and Illinois. To the northwest, the upper parts of the Paleozoic section have a broad, sagging synclinal trend that extends under Kansas and Nebraska. This sag has been called the Forest City basin. It is bordered on the west by a structure known as the Nemaha anticline, which extends from Kansas into Nebraska.

In addition to the Ozark dome, there are many other definite folds in younger rocks in various parts of the state. These are primarily elongate anticlinal structures that generally have a northwest to southeast trend. Some of these can be recognized on the geologic map as the elongated, northwest-southeast outcrops of older rocks surrounded by younger. Prominent among these is the Lincoln fold, which extends from Lincoln County northwest to Ralls and on into Knox and Scotland counties. Here Ordovician outcrops are surrounded by Silurian and Devonian beds, which in turn are surrounded by Mississippian rocks. Another is the Saline County arch in Cooper and Saline counties, which can be followed as far as Livingston County. Another is the Brown's Station anticline in northeastern Boone County, where Mississippian beds are surrounded by younger Pennsylvanian rocks. This anticline is paralleled by the Auxvasse Creek anticline in Callaway and southwestern Audrain counties. In southwestern Montgomery County the Mineola dome is well exposed in road cuts along Interstate 70 where the highway crosses the valley of the Loutre River. Here the Ordovician beds are surrounded by exposures of Silurian, Devonian, and Mississippian rocks.

A circular structure in eastern St. Louis County has been named the Florissant dome. It contains about one hundred feet of closure (that is, it is about one hundred feet vertically from top to bottom of the dome) in the Kimmswick Dolomite, and this formation has produced oil since 1953. St. Peter Sandstone is also involved in this structure and because of its porosity is used for underground gas storage. The sandstone is surrounded above and below by impermeable beds that prevent the gas from escaping.

Many other similar structures cannot be seen on the map in Plate 3 because of the lack of detailed distinction between some of the beds. Mary McCracken (1971) has reported that there are nearly one hundred named anticlines in Missouri, and she locates and describes most of them. This abundance has caused some geologists to refer to the "wrinkled skin" of Missouri.

The pre-Pennsylvanian rocks of the state are broken by many faults, but they are too small to show on the geologic map. The area east and north of the St. Francois Mountains is intensively faulted, primarily in a northwest-southeast trend. This is also true of the southwestern part of the state in the Springfield Plateau region. A geologic map of the state at a scale of 1 to 500,000 (about 1 inch = 8 miles) shows the faults very clearly. It is beyond the scope of this book to elaborate on all the individual structures of Missouri, but McCracken's *Structural Features of Missouri* gives detailed descriptions of many.

Joints are another common feature of the consolidated rocks, which are more rigid and less flexible. They are fractures formed in response to pressure or stress. Unlike folds or faults, joints do not involve horizontal or vertical movement. Joints can be seen on the top surfaces of layers of rock, called the bedding planes, where they usually occur in groups or sets and form rectangular or rhomboid blocks (Figs. 4–6 and 4–7). McCracken points out that nearly all of the consolidated rock formations of Missouri are jointed. In general, the joints are vertical with one side having a northwest direction and the other a northeast. These joint systems create openings through which ground water can move, thus allowing for the creation of caves and springs.

The faults and folds described by McCracken are in general spread widely across the state. However, six aberrant features whose origins are not clearly understood occur along a line across southern Missouri. They have been described by Frank Snyder and Paul Gerdeman (1965) as "intensely disturbed areas variously interpreted as cryptovolcanic structures, cryptoexplosive structures, or meteorite-impact scars." (*Crypto* means covered, hidden, latent, or secret. *Cryptovolcanic* and *cryptoexplosive* are used to designate rock structures that are nearly circular in shape, relatively local in size, and that show deformation similar to that

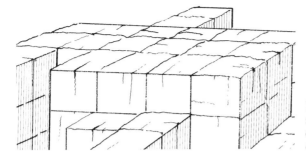

Fig. 4–6. Block diagram showing the development of joints in massive, firmly cemented rocks.

Fig. 4–7. The weathered surface of Pennsylvanian limestone showing joint patterns on the bedding plane. DNR–Vineyard.

caused by volcanic or other explosive activity, but whose origins are problematic or in question.) The names of the six Missouri features are, from east to west, Avon diatremes, Furnace Creek volcanics, Crooked Creek structure, Hazelgreen volcanics, Decaturville dome, and Weaubleau disturbance. A description of each one in turn will help to explain what they are (Fig. 4–8).

Avon diatremes

In western Ste. Genevieve and eastern St. Francois counties the Upper Cambrian rocks contain streaks of igneous rock and other intensely

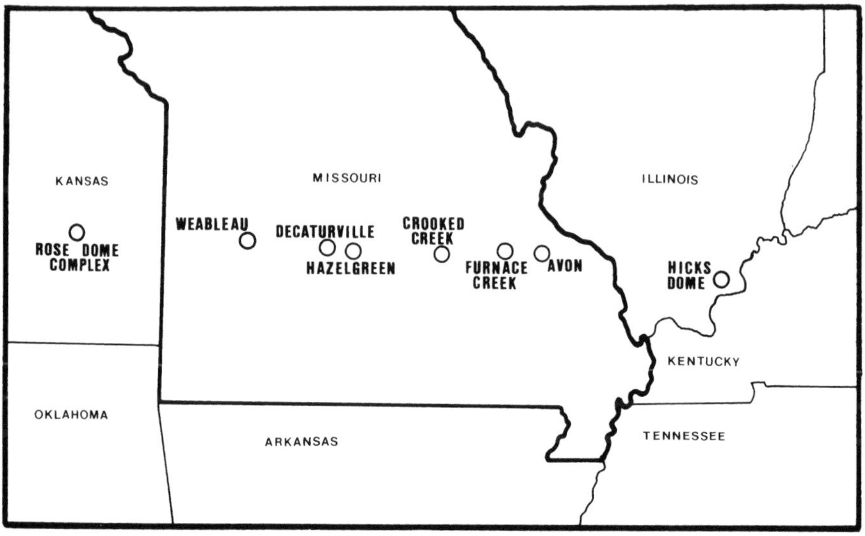

Fig. 4–8. Cryptoexplosive and cryptovolcanic structures across southern Missouri. From Snyder and Williams et al., 1965.

shattered igneous material that appear to have been caused by explosions. Also mixed with the igneous fragments of Precambrian age are blocks of sedimentary formations that are as young as Devonian. Snyder and Gerdeman place the time of formation of these structures in the Late Devonian Period.

Furnace Creek volcanics

The Upper Cambrian Lamotte Sandstone of Washington County contains a funnel-shaped crater filled with material resembling that commonly ejected from explosive volcanoes. And in the lower part of the Bonneterre Formation there is a layer of volcanic ash that was deposited at the same time as the lower part of the Bonneterre Formation. The lower part of the funnel extends downward into Precambrian granite.

Crooked Creek structure

In Crawford County there is a complex circular, ringlike structure of Cambrian and Ordovician formations that measures about three by four miles. Here the oldest formation exposed is the Bonneterre, which occurs about 1,000 feet higher than its normal position for the area. The youngest formation involved in the structure is the Jefferson City, so the disturbance must have come after the Jefferson City was deposited. This structure is at the intersection of major faults and is most likely volcanic in origin.

Hazelgreen volcanics

North of Hazelgreen in Laclede County a drill core in the Lamotte Sandstone revealed a thirty-five-foot section of volcanic ash and other volcanic detritus. The source of this material is thought to have been about two or three miles from the location of the drill hole, but no volcanic sites have been found.

Decaturville dome

In Laclede and Camden counties the Decaturville dome is a wooded ring of low hills about three and a half miles across. In general it consists of an uplifted, intensely broken and crushed core approximately one mile in diameter surrounded by an outer zone of broken and down-faulted younger formations that can be identified as Cambrian and Ordovician dolomites. It is sharply fault-bounded and shows intense and unusual deformation. The rock in the center has a coarse granite-like texture and resembles the Precambrian granite of the St. Francois Mountains.

This structure has been looked at by many geologists over many years and variously described as cryptovolcanic or of impact origin. A detailed study by T. W. Offield and H. A. Pohn (1979), used drill cores, small-scale mapping, and meticulous mineral examination of the surrounding Cambrian and Ordovician formations. Blocks of these formations have been uplifted as far as 1,500 feet above normal position. Many blocks are deformed and brecciated. Within the structure, blocks of the Derby–Doe Run dolomite bear well-formed shatter cones. The degree of deformation and displacement of the formations, the ringlike structure and accompanying shatter cones, plus intergranular alterations in some of the minerals, lead these workers to believe that this structure was caused by a meteor or comet impact. The age is post-Pennsylvanian, perhaps as young as Cretaceous.

Weaubleau disturbance

A thirty-square-mile area in St. Clair County is partly covered by un-disturbed Pennsylvanian sediments within which earlier Jefferson City and Kinderhook beds from the Mississippian Period are intensely faulted and broken up. In some localities, Ordovician beds are interthrusted into Mississippian beds. The Mississippian rocks also contain conglomerates not known elsewhere in rocks of this age. There are no "ring structures" here as at Decaturville and Crooked Creek, but this structure is also interpreted as being explosive in origin. It is thought to have been caused by a buildup of gas pressure that was released slowly in early Mississippian (Kinderhook) time but became violently explosive during the later Mississippian Burlington time. The area was later covered by Pennsylvanian sediments.

*　　*　　*

Analyzing the structure of Missouri tells us that there have been several major periods of crustal disturbance, and by studying the ages of the rocks we can put these in chronological order. The first and probably the most severe disturbance happened during Precambrian time and was accompanied by the outflowing of the extrusive igneous and volcanic materials we find today and by the emplacement of the deep-seated intrusive masses. Even though the Precambrian rocks are exposed only in the St. Francois Mountains, their variety, their complex faulting, and other relationships tell us that there were different periods of folding, faulting, and magmatic movement.

McCracken (1971) has analyzed structures and unconformities throughout the state to define at least five deformational events in addition to this Precambrian activity. The first of these was in late Ordovician time, and it created an unconformity between the Canadian and Champlainian formations. The second came after early Devonian time but before Mississippian. In some places it might have begun in Late to Middle Devonian. The third came after the Mississippian but before the Middle Pennsylvanian. The fourth was very widespread and closed out the Pennsylvanian. It left remnants of Pennsylvanian sediments in contact with Ordovician dolomites in many places throughout the area between the Missouri River and the northwestern part of the Ozark–St. Francois region. The last pronounced deformation happened in Tertiary time after the close of the Paleocene Epoch but before the Pliocene. The area most affected by this action is the Mississippi Embayment.

New Madrid and the Earthquakes

At two o'clock on the morning of December 16, 1811, the residents of New Madrid, Missouri, and the surrounding area were shaken from their

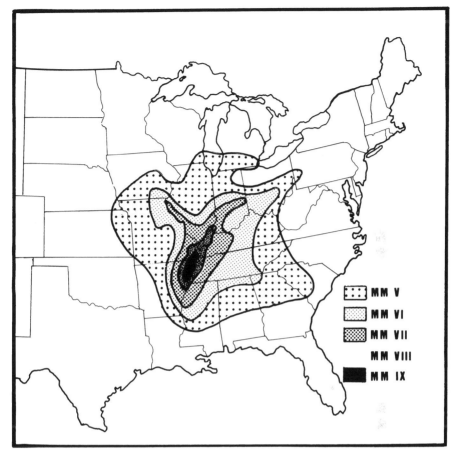

Fig. 4–9. Shaded areas represent possible intensity levels of an earthquake of magnitude 6.7 in the New Madrid seismic zone. (MM = Modified Mercalli Scale.) DNR.

sleep by the first in a series of earthquake shocks that lasted more than a year. New Madrid was completely destroyed, and the effects of the quake were felt over an area that reached as far as New Orleans, Charleston, Savannah, Baltimore, Pittsburgh, and Richmond, Virginia. In Washington, D.C., church bells rang and paved streets cracked (Fig. 4–9).[1]

At that time, there had been almost no research or scientific observation of earthquakes. The seismograph had not yet been invented. There is no instrumental record of the strength of this quake, but using reports from newspapers, magazines, personal letters, and journals of the time, seismologists have estimated that this early-morning quake would rank at least 8.0 on the Richter Scale, as serious as any ever felt in what is now the United States.

This early-morning shock was only the first of a series of strong shocks that continued into early 1812 and was followed by other weaker tremors later in the year. One observer in Louisville, using homemade pendulums, recorded 1,874 shocks. The most intense were those of December 16, 1811 (the first), January 23, 1812, and February 7, 1812. Based on reports of damage to buildings and other indirect evidence, Otto Nuttli (1973) has estimated their Richter Scale magnitudes at 8.6, 8.4, and 8.7, respectively.

Confusion arises in the discussion of earthquakes because the terms *intensity* and *magnitude* are not synonymous. Intensity is a measure of the effects of an earthquake at a particular place, taking into account the geologic structure of the affected area, the depth of focus, and the distance from the epicenter. Intensity is measured on the Modified Mercalli Scale, which has twelve degrees ranging from I (usually detected only by instruments) to XII (causing major to total damage). Magnitude is measured on the more familiar Richter Scale and is related to the energy released at the focus of the quake. It is expressed as a number from 1 to 10. The strongest earthquakes in recorded history have ranged in magnitude from 8.4 to 8.6. It is important to understand that intensity increases by a factor of ten for every number on the scale. That is, an earthquake measuring 3.0 is ten times as strong as one measuring 2.0, one at 4.0 is ten times as strong as one at 3.0, and so on, so that an earthquake of 8.0 is 10 million times as strong as one measuring 1.0 (see Table 4–1).

Newspaper and magazine reports described the devastation from the New Madrid quakes—buildings of brick, stone, and logs were demolished. Stone buildings as far away as St. Louis and even Savannah, Georgia, were cracked. Large expanses of land were flooded; acres of timber were submerged; boats were swamped and broken up; riverbanks slumped into the river; islands disappeared and new ones emerged. The channel of the Mississippi River was altered. Sloughs were drained; others filled to become lakes. One example often cited is the birth of Reelfoot Lake, in Tennessee, where the area subsided to form a lake at first reported to be more than fifty feet deep. Later, more reliable reports, cited by James Lal Penick (1981), gave the depth as twenty-five feet. Geyserlike eruptions of water carrying sand and coal were widespread in the lowlands of the river valley. The Mississippi is reported to have temporarily reversed its flow. The site of the town of New Madrid settled from twenty-five feet above river level to only twelve feet.

There is no precise knowledge of the loss of human lives. There were drownings as boats were swamped and as the riverbanks collapsed. Many homes were destroyed by the flowing waters, and the occupants probably drowned. Reports at the time claimed strange behavior among dogs, horses, and birds. Wild animals of various kinds were reported to have sought the company of humans, apparently seeking protection from the strange noises and the quivering of the land.

The behavior of the people was no less bizarre than that of the animals. Some became faint and nauseous. Some reported aches and pains in their joints. Many were panic-stricken to the point of temporary insanity. Most feared that the quakes were heaven-sent, and preachers flocked to the region. One of them, Reverend James P. Finley, states in his autobiography that during a daylight shock on December 16, 1811, "Consternation sat on every countenance, especially upon the wicked. . . . It was a time of great terror to sinners."[2] Throughout the area where the quakes were felt most strongly, church membership grew. The Methodist Church grew by 50 percent from 1811 into 1812. Many blamed the shocks on the appearance of Halley's comet in 1811, which was generally visible in the United States from September to mid-January of 1812. The Indians, who outnumbered the white settlers, were convinced that the Great Spirit was angry. The naturalist John James Audubon, traveling through the region on horseback in 1812, experienced one of the major quakes and recorded the event in his journal.

Indian lore and indisputable geologic evidence demonstrate that the 1811–1812 quakes were not the first to shake the region. Myron Fuller (1912) refers to a report by Daniel Drake of shocks felt in 1776 in Ohio, in 1791 or 1792 in northern and northeastern Kentucky, and in 1795 in the Illinois Territory.[3] Penick (1981) mentions an earlier report of a quake on Christmas Day 1699 near the site of Memphis.

Records have been maintained since 1909, when the first seismograph in the Midwest was installed at St. Louis University. The most noticeable recorded quakes were on October 21, 1965; November 9, 1968; and March 25, 1976. The last was centered near Marked Tree, Arkansas, and was felt over seven states. It was ranked at magnitude 5.0 on the Richter Scale.

In 1974 St. Louis University and the U.S. Geological Survey installed a network of modern earthquake-recording instruments in the seismic zone. In 1979 Memphis State University, with support from the Nuclear Regulatory Commission, set up another network to augment the earlier one. Between mid-1974 and mid-1981 these instruments detected an average of 150 events per year with magnitude greater than 1.0. Between 1974 and 1985 more than 2,000 quakes were detected by instruments, but 97 percent were not felt (Fig. 4–10). Magnitude/frequency studies and other data indicate that large earthquakes in the

TABLE 4-1

Comparison of Modified Mercalli Seismic Intensity Scale and Richter Seismic Magnitude Scale

Modified Mercalli Seismic Intensity Scale	Richter Seismic Magnitude Scale
I. Not felt. Marginal and long-period effects of large earthquakes.	3.5
II. Felt by persons at rest, on upper floors, or favorably placed.	4.2
III. Felt indoors. Hanging objects swing. Vibration like passing of light trucks. Duration estimated. May not be recognized as an earthquake.	4.3
IV. Hanging objects swing. Vibration like passing of heavy trucks; or sensation of a jolt like a heavy ball striking the walls. Standing motor cars rock. Windows, dishes, doors rattle. Glasses clink. Crockery clashes. Wooden walls and frames creak.	4.8
V. Felt outdoors; direction estimated. Sleepers wakened. Liquids disturbed, some spilled. Small unstable objects displaced or upset. Doors swing, close, open. Shutters, pictures move. Pendulum clocks stop, start, change rate.	4.9–5.4
VI. Felt by all. Many frightened and run outdoors.* People walk unsteadily. Windows, dishes, glassware broken. Knickknacks, books, etc., off shelves. Pictures off walls. Furniture moved or overturned. Weak plaster and masonry D (weak masonry) cracked. Small bells ring (church, school). Trees, bushes shaken visibly, or heard to rustle.	5.5–6.1
VII. Difficult to stand. Noticed by drivers of motor cars. Hanging objects quiver. Furniture broken. Damage to masonry D, including cracks. Weak chimneys broken at roof line. Fall of plaster, loose bricks, stones, tiles, cornices, unbraced parapets, and architectural ornaments. Some cracks in masonry C (ordinary masonry). Waves on ponds; water turbid with mud.* Small slides and caving in along sand or gravel banks.* Large bells ring. Concrete irrigation ditches damaged.*	6.2 to 6.9

VIII. Steering of motor cars affected. Damage to masonry C; partial collapse of masonry D. Some damage to masonry B (good masonry); none to masonry A (excellent masonry). Fall of stucco and some masonry walls. Twisting, fall of chimneys, factory smokestacks, monuments, towers, elevated tanks. Frame houses moved on foundations if not bolted down; loose panel walls thrown out. Decayed piling broken off.* Branches broken from trees.* Changes in flow or temperature of springs and wells.* Cracks in wet ground and on steep slopes.*   6.2 to 6.9

IX. General panic.* Masonry D destroyed; masonry C heavily damaged, sometimes with complete collapse; masonry B seriously damaged. General damage to foundations. Frame structures shifted off foundations, if not bolted. Frames cracked. Serious damage to reservoirs.* Underground pipes broken.* Conspicuous cracks in ground.* In alluviated areas sand and mud ejected, earthquake fountains, sand craters.*   6.2 to 6.9

X. Most masonry and frame structures destroyed with their foundations. Some well-built wooden structures and bridges destroyed.* Serious damage to dams, dikes, embankments.* Large landslides.* Water thrown on banks of canals, rivers, lakes, etc.* Sand and mud shifted horizontally on beaches and flat lands.* Rails bent slightly.*   7.1–7.3

XI. Rails bent greatly.* Underground pipelines completely out of service.*   7.4–8.1

XII. Damage nearly total. Large rock masses displaced.* Lines of sight and level distorted.* Objects thrown into the air.*   8 or more

These criteria may be misleading as a measure of the strength of shaking.

	Richter Relative Magnitude	
	Scale No.	Strength
Intensity—A measure of the effects of an earthquake at a particular place, taking into account the geologic structure of the affected area, the depth of focus and distance from the epicenter.	1	1
	2	10
	3	100
	4	1,000
Magnitude—A measure of the energy released at the focus of the quake.	5	10,000
	6	100,000
	7	1,000,000
	8	10,000,000

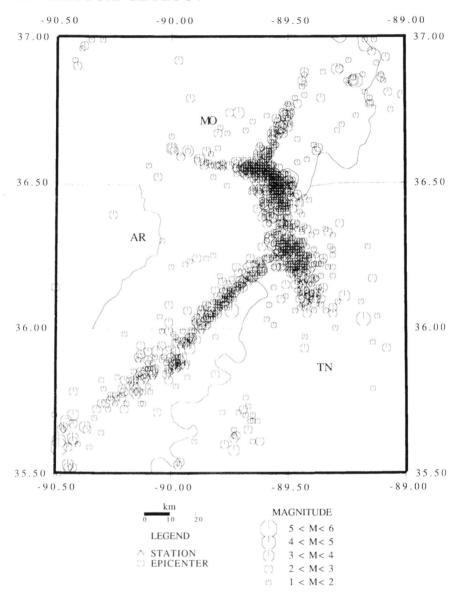

Fig. 4–10. Seismicity map of the New Madrid area showing the frequency of earth-quakes from June 29, 1974, through June 27, 1990. Courtesy Brian J. Mitchell, St. Louis University.

New Madrid region can be expected to recur every six to seven hundred years.[4] Robert M. Hamilton and A. C. Johnston (1990) estimate that the probability of an earthquake of magnitude 8.0 in the area during the next fifty years is low, in the range of 2.7 to 11 percent, and that the probability of a 6.0 to 6.5 earthquake is 45 to 97 percent. Nonetheless, as Nuttli (1973) has noted, the next large one "might occur as soon as next year or as late as several thousand years."

In December 1989 Iben Browning, a "self-taught climatologist," predicted that an earthquake would occur in the New Madrid area on December 3, 1990. Seismologists did not take his prediction seriously. As could be expected the news media became excited and prepared many stories on past earthquakes and the potential destructiveness of a future earthquake. On the predicted date the little town of New Madrid was crowded with reporters and cameramen. Even the governor and a Missouri senator were on hand. Nothing happened. Later reports, outlined by Sue Hubbell (1991), said that the crowd of visitors and news reporters loaded the city so heavily that the earth could not possibly have budged.

In 1811–1812, the area most severely affected by quakes was only sparsely populated with no cities of appreciable size. If the same area were to be hit again by a similar quake, or series of quakes, the effects would be devastating. The area now contains St. Louis, Memphis, Evansville, Louisville, Cape Girardeau, and other cities whose total population is estimated at approximately 12 million. A. C. Johnston (1982) calculated that if another quake equal in strength to those in 1811–1812 were to strike in the year A.D. 2000, more than 3,000 persons would be killed, and property damage in Memphis alone would be more than $1.3 billion. Nuttli (1973) argued that earthquakes in the New Madrid seismic zone may cause damage as much as one hundred times greater than those of similar magnitude in western North America. The reason is that the rock structure of the earth's outer crust in the western United States tends to "soak up" the shock waves, while in the central and eastern area this happens to a far lesser degree.

The New Madrid seismic zone is very complex structurally, and the geology is not well understood. The area seems to be the location of an intracontinental rift that developed in Late Precambrian and Early Cambrian time. Because of the thick cover of later sediments over the earthquake-causing structures, details are difficult to determine. Seismic reflection studies suggest a number of cross-cutting structural features, which constitute a complex of faults. Studies of microquakes in the area since 1974 indicate that these faults are mainly in the granitic basement.

Johnston (1982) has postulated the presence of three major, deeply buried faults in the region. One strikes southeast from near New Madrid into northwestern Tennessee. Another, longer one extends about one hundred kilometers southwest out of Missouri into northeastern Arkansas. A third, which is less well developed, extends northward from the first one to the vicinity of Cairo, Illinois. Johnston's discussion of the geology of the New Madrid area raises the question of why these earthquakes occur in the Midcontinent region instead of near the edges of colliding plates. The answer seems to be that this area was, in Mesozoic time, covered by a northward extension of the Gulf of Mexico in which thick layers of fine-grained sediments accumulated and that they are still not well consolidated.

An interesting sidelight on history is recalled in Hamilton and Johnston's "Tecumseh's Prophecy: Preparing for the Next New Madrid Earthquake" (1990). Penick (1981) recounts the legend that Shawnee Indian chief Tecumseh predicted the first of the New Madrid earthquakes of 1811–1812. Hamilton and Johnston cite Tecumseh's warning, not as a documented prediction of the earthquake, but as an interesting lead-in to their recommendations for geologic and geophysical studies in the New Madrid region. Further studies may lead not only to better ways of predicting future quakes but also to ways of reducing losses to lives and property when the inevitable happens. Continued observations of the quakes of this area, large and small, will result in greater understanding of the geology of the area and ultimately may help scientists develop technology for forecasting future shocks.

5

KARST

The most abundant rocks in Missouri are the carbonate rocks lime-stone and dolomite. As can be seen on the geologic map in Plate 3, these beds are at, or near, the surface over a large part of the state. In addition to being of economic importance as construction materials and as the host rocks for metallic ores, these beds serve as water-bearing formations, or aquifers, and provide many beautiful springs and in-triguing caves. A less desirable feature is that the same characteristics that allow for the formation of caves and springs also lead to the cre-ation of sinkholes. In fact, caves, springs, and sinkholes are closely inter-related and occur throughout the carbonate region of Missouri.

The carbonate rocks are predominantly composed of the minerals cal-cite ($CaCO_3$) and dolomite ($CaMg(CO_3)_2$); both, especially calcite, are soluble in acidic ground water. When rain falls through the air, it absorbs carbon dioxide and thus becomes acidic. As it percolates through the soil, it becomes more so. When the water reaches carbonate bedrock, it dis-solves the minerals in the rock, bringing them into solution. Most lime-stones and dolomites are porous and contain crevices, joints, or other openings through which the water can move. By the process of solution, these openings become enlarged as the dissolved material is carried away. Given eons of time, these openings grow to form extensive cave passages and voids of many sizes. In some localities, elaborate underground drain-age systems develop. Springs are the outflow from these underground drainage systems (Fig. 5-1). Areas where caves, springs, and sinkholes are found, that is, where rocks exist that are subject to dissolution, are called *karst areas*. The name is derived from a region in Yugoslavia, near the Adriatic Sea, where such phenomena were first studied.

Missouri ranks high on any list of karst regions. There are major karst areas in the Mississippian rocks of St. Louis, Ste. Genevieve, Cooper, Greene, Boone, and Christian counties, and in Ordovician rocks of Per-ry, Phelps, Pulaski, and Howell counties. Karst development sometimes creates a complex assortment of caves, tunnels, bridges, and arches in a

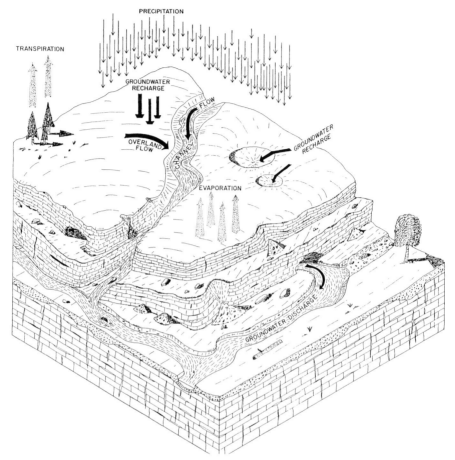

Fig. 5–1. The hydrologic cycle of a Karst region leading to cave and sinkhole formation. DNR–James E. Vandike.

relatively small area. Examples are the area in Oregon County called Grand Gulf and the Ha Ha Tonka area in Camden County along a southern arm of the Lake of the Ozarks. Both of these karst complexes are now state parks.

Caves

Missouri is widely known as "The Cave State," and the name is well deserved. Caves are found in many regions of the state, with the exception of the areas north of the Missouri River where Pennsylvanian rocks are covered by glacial drift. The greatest concentration is in Perry County, which has 630 known caves. The longest caves are also in Perry County (see Table 5-1). Crevice Cave, at 28.2 miles, is not only

TABLE 5-1

Lengths of Missouri Caves

Longest Caves as Known in 1986

	Feet	*Miles*	*Kilometers*
Crevice Cave, Perry Co.	148,902	28.20	45.37
Moore Cave System, Perry Co.	102,362	19.38	31.18
Mystery Cave, Perry Co.	89,760	17.00	27.35
Rimstone River Cave, Perry Co.	74,110	14.04	22.59
Carroll Cave, Camden Co.	59,400	11.25	18.10
Devil's Ice Box, Boone Co.	30,421	5.76	9.27
Piquet Cave, Pulaski Co.	25,637	4.86	7.82
Cameron Cave, Marion Co.	24,395	4.62	7.43
Hot Caverns, Perry Co.	15,790	2.99	4.81
Cathedral Cave, Crawford Co.	15,790	2.99	4.81

Longest Show Caves

	Feet	*Miles*	*Kilometers*
Cameron Cave, Marion Co.	24,395	4.62	7.43
Mark Twain Cave, Marion Co.	11,300	2.14	3.44
Meramec Caverns, Franklin Co.	10,900	2.06	3.31
Onondaga Cave, Crawford Co.	9,106	1.72	2.77
Ozark Underground Laboratory, Taney Co.	9,025	1.71	2.76
Fantastic Caverns, Greene Co.	6,715	1.27	2.04
Marvel Cave, Stone Co.	6,703	1.27	2.04
Round Spring Caverns, Shannon Co.	6,510	1.23	1.98
Fisher Cave, Franklin Co.	6,250	1.18	1.90
Ozark Caverns, Camden Co.	3,400	.64	1.03

the longest cave in Missouri but also the eighth longest in the United States.[1]

As of 1991, 5,100 caves were on record at the Division of Geology and Land Survey. Thanks to the efforts of cave-exploring groups cooperating with the Missouri Geological Survey, many of these caves have been mapped. In the mid–1950s, State Geologist Thomas R. Beveridge worked out an agreement with a group of cavers whereby the Geological Survey (now part of the Department of Natural Resources, Division of Geology and Land Survey) became a repository for cave maps and other data. Since then, more than 2,000 cave maps have been deposited with the state agency. Many of these maps have been published in reports, journals, and books; copies are available through a reproduction service provided by the agency.

Caves occur in a wide variety of patterns, and these are largely controlled by the rock structure. Table 5–2 provides estimates of the ages and types of the rocks that serve as host formations for Missouri's caves. Jointing in many rock formations creates intersecting passageways that develop into mazelike arrangements of corridors and cross channels (Fig. 5–2). In less jointed beds, the patterns may resemble meandering streams with branching tributaries. Examples are Pleasant Valley Cave in Jefferson County and Onondaga Cave in Crawford County.[2]

Everyone knows what a cave is, but the term is sometimes applied loosely to overhanging rock ledges or hollow shelters. J. Harlen Bretz (1956) defines a cave as "a natural roofed cavity in a rock which may be penetrated for an appreciable distance by a human." Many caves are deep below the general land surface, some as far as 200 or 300 feet. In fact, the deepest known cave in the state is 383 feet deep. Others are at shallower depths. The latter, of course, have thinner "roofs" and are more subject to collapse. Such a collapse may leave segments of the cave open to the surface, resulting in a steep-walled surface stream or, if there is no stream in the cave, a dry, open exposure of the cave. In many localities in Missouri the collapse has been incomplete, and tunnels, natural arches (Fig. 5–3), and natural bridges (Plate 8) are the result. Beveridge (1978) has described twenty-three arches and twenty-five bridges. The difference between the two is more or less a matter of scale. Where the noncollapsed segment is of considerable length, the term *tunnel* is applied, and Beveridge describes twenty of these.

The caves of Missouri have been used by man from the earliest days (Fig. 5–4). Primitive human remains have been found in twenty-nine Missouri caves, and archaeologists have found many artifacts and other evidence that caves were occupied by early man and by the later Indians.

TABLE 5-2

Distribution of Missouri Caves by Rock Type and Age

Rock type	Number of caves (estimated)
Dolomite	3,110
Limestone	2,035
Sandstone	50
Igneous	5

Age	Number of caves (estimated)
PALEOZOIC ERA	
Pennsylvanian	75
Mississippian	1,475
Devonian	50
Silurian	??
Ordovician	3,200
Cambrian	400
PRECAMBRIAN ERA	5

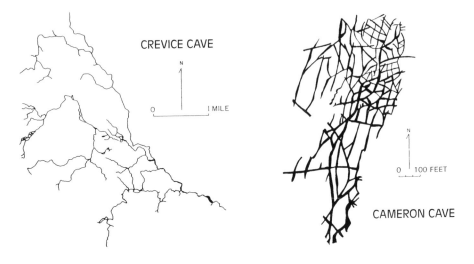

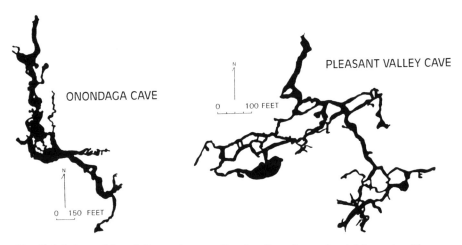

Fig. 5–2. Maps of four Missouri caves. Crevice Cave has a dendritic, or treelike, pattern with unusually long passageways (note that the scale for this cave is in miles, while the others are in feet). Cameron Cave is an example of a cave in which the pattern is created by intersecting joints. Onondaga and Pleasant Valley caves both have meandering stream-like patterns.

Fig. 5–3. The Pinnacles, north of Columbia, has several natural arches. DNR–Vineyard.

Even before man, the caves served as animal shelters. Cave explorers have found fossil remains of many Ice Age animals, including musk-ox, mastodon, armadillo, alligator, lion, saber-toothed tiger, bear, camel, sloth, and peccary.

In historic times caves have served many purposes and, as one would expect, there are many stories and legends, some true, about caves. Some stories deal with early French and Spanish explorers, and it is not surprising to find that many relate to hidden, and lost, treasures of gold and silver. Early trappers used the underground shelters for fur storage in the late 1700s. At least as early as 1840, the brewers in St. Louis used the numerous caves beneath the city as natural, ideal storage chambers. At least twenty-eight caves are known to be under the city of St. Louis, and several were used in the late 1800s as underground saloons and beer gardens.

Fig. 5–4. Graham Cave, a large, archlike opening in the St. Peter Sandstone, was used as a dwelling place by early cave men. The cave is north of Interstate 70, east of Loutre River, near Danville. DNR–Vineyard.

During the Civil War caves were used to conceal munitions and gunpowder and as sources for saltpeter used in making explosives. They also became hiding places for guerrilla bands, in addition to serving as "stations" on the "underground railroad." Tradition tells us that after the Civil War, outlaw gangs and marauders used caves as hiding places for themselves and for their loot. Among these were the Daltons, the Doolins, and the Youngers. Of course, the most famous of these outlaws was Jesse James who, if we believe all the tales, must have been in every cave in the state. Meramec Caverns, at Stanton, makes a strong claim to having been used by James and his gang.

In the later 1800s, caves became popular centers for recreation. In many cases their close association with the large springs and scenic rivers resulted in the development of parks, camps, and other types of recreation facilities, many of which are still in operation today. Also during this period, and on into the next century, cave onyx became a popular item for jewelry and ornamental stone. Mushroom "farming" was tried from the late nineteenth to the early twentieth century but was never a big success.

New uses were devised for the underground spaces in the twentieth century. National prohibition in 1910 "drove those who wished to make alcoholic beverages underground."[3] There are many stories of bootleggers and moonshiners and their secret hiding places. Fantastic Caverns, once known as Knox Cave, was owned by the Ku Klux Klan from 1924 to 1930. During World War II, caves were used in civil defense efforts as storage vaults for emergency food and water supplies. Plans were also made for them to serve as bomb shelters if needed. The U.S. Navy even drew up plans for building a secret jet propulsion lab in Onyx Cave near Waynesville, which has a huge, high-ceilinged room.

Mark Twain's stories about Tom Sawyer have brought fame to the Hannibal region. Today the cave where, in fiction, Tom and Becky were lost and then rescued is known as Mark Twain Cave. In Twain's account he called it McDougal's Cave. The real owner was Dr. Joseph Nash McDowell. The pattern of this cave is more strongly related to jointing than is true for most Missouri caves, and it differs as well in being almost free of the cave features called *speleothems*, that is, the stalactites and other cave mineral deposits commonly associated with caves.

These spectacular speleothems are one of the main attractions in most caves (Plate 9). They are created as water seeping from the rock loses its carbon dioxide or evaporates and leaves behind the minerals it had carried in solution. Speleothems formed from flowing water are called *flowstone,* and those created by dripping water are called *dripstone.* Those that hang from the ceiling are known as *stalactites.* They may form single cones, or several may coalesce to form complex shapes. In some cases the drip from the end of a stalactite may start an upward growth from the cave floor to form a *stalagmite.* If the stalactite and the stalagmite grow together, a *column* is formed. In some caves the drips from the ceiling may follow a curved crack, and the dripstone formed may take the shape of a pleated or draped curtain (Fig. 5–5).

The abundance of caves in Missouri has attracted a large number of cave enthusiasts. Those with formal scientific backgrounds are called *speleologists,* while those who are interested in exploration and mapping are called *cavers.* The Missouri Speleological Survey is a formal organization dedicated to the exploration, study, mapping, and conservation of Missouri caves.

More than a dozen caving clubs exist in Missouri, and most of them are affiliated with the National Speleological Society. The organization works to teach the public the pleasures, but also the dangers, of caving and to encourage the protection and preservation of caves. Many have been plundered and have suffered severe damage from vandalism. Approximately twenty-five caves are open to the public; the number varies from year to year and from season to season. Caves used as show caves, open to the public, must meet strict safety standards and are subject to approval by the state mine inspector. Anyone visiting a cave should

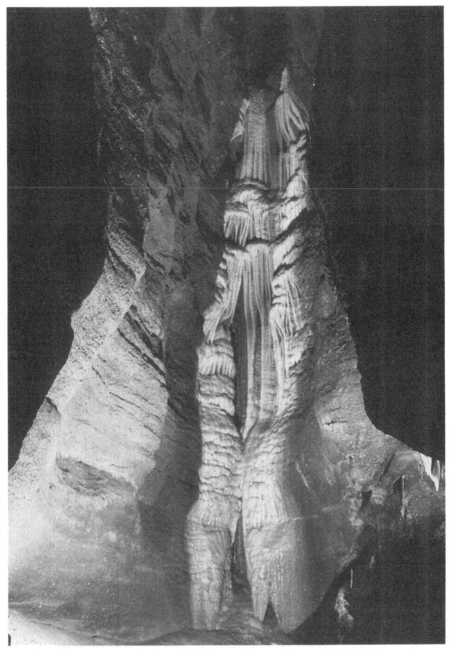

Fig.5-5. The beauty inside Devil's Icebox in Boone County. DNR–Vineyard.

remember that the sculptured rock forms and speleothems found in caves are nonrenewable and irreplaceable. Once damaged, they are damaged forever in our comprehension because they represent millions of years of geologic development.[4]

Springs

No one knows exactly how many springs are in the state, but the Division of Geology and Land Survey has recorded more than 1,100. Among these are some of the largest springs in the United States, even in the world. At least eleven have an average daily discharge of more than 50 million gallons each (Table 5–3). Thus, as Jerry D. Vineyard and Gerald L. Feder (1982) point out, "In an average day more than a billion gallons of water flow from the ten largest springs in Missouri." The largest, Big Spring in Carter County (Fig. 5–6), is known to have had a maximum one-day flow greater than one billion gallons.[5] The same spring has averaged a daily flow of 276 million gallons over a forty-nine-year period of record. The flow of the springs can be correlated with seasonal rainfall. The large volume of water flowing from these springs may seem strange and mysterious, but it should be remembered that the underground channels are complex networks and serve as drainage systems for the rainfall on many square miles of land surface (see Fig. 5–1).

Springs have long been recognized as priceless resources, and they have been important factors in the settlement and development of various parts of the state. The early settlers found them to be clean and attractive sources of water. In some places, especially Saline, Howard, and Perry counties, they provided the pioneers, and before them the Indians, with salt. In the Ozark area the strong flows provided water power to run grist mills, a few of which still operate on a limited basis. Some springs are highly mineralized and were believed to provide medicinal and health benefits. Spas and bath houses were popular for many years in the late nineteenth century (Fig. 5–7).[6] Today the major use of the springs is recreational, and many in their natural settings have become focal points for state and national parks and preserves. Some are used for the spawning and raising of trout and other fish for stocking the many beautiful streams of the Ozarks.

The quality of the water flowing from the springs is generally good from a mineralogical point of view. As would be expected, the water is "hard." That is, it carries a considerable load of dissolved minerals, which may vary with the season and with the amount of rainfall. Water from dolomites is heavy in magnesium, and that from limestones is heavy in

TABLE 5-3

Flow of Large Springs in Missouri
(In millions of gallons per day)

Name of Spring	County	Average Flow	Maximum Measured Flow
Big	Carter	276	840
Greer	Oregon	214	583
Double	Ozark	100[1]	150
Bennett	Dallas	100	([2])
Maramec	Phelps	96	420
Blue	Shannon	90[1]	153
Alley	Shannon	81	([2])
Welch	Shannon	75[1]	214
Boiling	Pulaski	68[1]	45
Blue	Oregon	61[1]	65
Montauk	Dent	53[1]	79
Hahatonka	Camden	48	123
North Fork	Ozark	—	49
Round	Shannon	26.5	336
Hodgson Mill	Ozark	24[1]	29

[1]*Estimated.*
[2]*Peak flows affected by runoff upstream from spring, after heavy rains.*

Source: Vineyard and Feder, 1982.

Date	Minimum Measured Flow	Date
June 1928	152	Oct. 6, 1956
May 26, 1927	67	Nov. 16–19, 1956
Apr. 7, 1965	30	Nov. 16, 1964
——	36	Nov. 13, 1934
1927–1928	36	Aug. 1, 1934
Apr. 24, 1964	40	Oct. 10, 1932
——	35	Oct. 1934
June 22, 1924	45	Aug. 24, 1964
Oct. 26, 1963	36	Jan. 21, 1964
July 18, 1935	35	Aug. 13, 1936
May 15, 1939	25	Aug. 13, 1934
June 19–20, 1924	28	Feb. 23, 1923
July 6, 1966	43	April 8, 1966
May 1933	6.5	Dec. 1937
Aug. 18, 1934	15	Aug. 29, 1923

Fig. 5–6. Big Spring, in Carter County, has the largest flow of any spring in Missouri. It rises from the Eminence Dolomite.

calcium. Other materials present are iron, manganese, sulfates, fluorides, and nitrates.

In 1892 Paul Schweitzer provided extensive coverage of Missouri's mineralized spring waters—their chemical constituents and uses for health and therapeutic purposes. He listed eighty-three localities in thirty-seven counties where samples were collected and analyzed. Several of these were localities where hotels, cabins, and bathing facilities were available. Others had facilities for shipping the water in barrels or bottles to "patients" at remote locations. Schweitzer believed, "Many of these waters indeed possess virtues of a very high order, which, in time, must render them famous beyond the confines of the State, and destine them to become sources of relief for suffering mankind." He made recommendations concerning drinking of, and bathing in, the several "kinds" of waters available in Missouri. The supposed powers of the various springs ranged widely but included cures and beneficial effects for rheumatism, kidney diseases, stomach troubles, blood diseases, cold sores, malaria, heartburn, dyspepsia, Bright's disease, dropsy, eczema, and liver disorders. Some were recommended for "strengthening enfeebled constitutions" and curing "general debility." One spring was even named Cure All Spring.

Fig. 5-7. Sweet Spring in Saline County was so-called because the water that flowed from the limestone lacked the salty or strong mineral taste of most springs in the area. Its flow of eleven hundred gallons per hour was used for both bathing and drinking. Facilities at the spring included a "well-appointed" hotel of 130 rooms, a ballroom and pavilions, and other structures including the spring house illustrated here. This was only one of many spas built in Missouri during the nineteenth century.

In most springs the water is clear and colorless, but during certain seasons it may become pale brown with dissolved organic compounds. The temperature of spring water is remarkably constant year-round and stays close to the mean annual surface temperature of 58 to 59 degrees F.

Because of their constant cool temperature, springs contain distinctive plant and animal life. J. A. Steyermark (1941) reported sixty species of seed-producing plants from springs and spring-fed branches. Sixteen of these were common in most springs. The most abundant of these is watercress, but other common species are water milfoil, water starwort, and waterweed. Bryophytes and algae are also common.

Vineyard and Feder (1982) include a section by Wm. L. Pflieger describing the fauna of the springs. Though limited in comparison with streams and lakes of the region, the fauna is more diverse than might be expected. The principal vertebrates are fish and salamanders, with some frogs. The invertebrates are flatworms, crustacea, insects, and snails. Of particular interest are two species of crayfish that over many generations have adapted to the dark underground waters and become blind and white. Two species of blind, white fish also inhabit Missouri springs, but they are very rare. Blind, white salamanders are more common.

Although spring water may be clear, cool, and colorless, it can still be polluted or contaminated and should never be used for drinking or other human consumption or use without treatment, in spite of Schweitzer's recommendations. Many people fail to realize that springs are recharged from rainfall, and surface water can become polluted by septic tanks, agricultural practices, or spills, accidents, and other contaminant sources. Although some of the early settlements were at or near a spring, this source of water for municipal consumption is very limited. Currently, only the city of Springfield and the town of Mill Spring obtain part of their water supply from springs.

There are springs in many parts of the state, but their location and distribution are controlled by the geologic character and structure of the rocks. As can be seen on Plate 1, the majority of the springs in the state are in the Ozarks region. The Cambrian and Ordovician dolomites of this region are readily soluble and easily fractured and broken, expediting the movement of the water. Here the massive beds of nearly 2,000 feet of dolomite and associated sandstones provide colossal underground storage reservoirs for the rainfall that finds its way through the thin soils to the passageways and ultimately issues as springs. Recreation parks, fish hatcheries, and cattle farms are the greatest users of the springs in this area.

The Springfield Plateau is the second most prolific spring area. Here the underlying beds are Mississippian limestones and the most abundant chemical in the water is calcium carbonate. Other conditions similar to the Ozarks are present, except that the springs are not as large and their flow is less constant. The volume of reservoir storage is, of course, much less.

A few springs are found in the plains of northern Missouri. Because these till plains are underlain mostly by Pennsylvanian shales and coals, there are not many localities where springs develop, but a few issue from Pennsylvanian limestones.

*   *   *

Although caves and springs have been discussed separately, in many locations a cave and a spring are one and the same. In some of these the spring is the surface exposure of an underground stream. In others it is the site where the spring becomes the "head" of a surface stream. Divers have explored several springs and probed the extensive underground streams that flow through the water-filled passages.[7] In fact, southern

Fig. 5-8. Sinkholes in the Florissant Karst area in east-central Missouri.
DNR-Vineyard.

Missouri is a veritable cave factory where one can see caves in all stages
of development, from the completely water-filled cavities of springs to
"completed" caves with extensive speleothems and large passages where
streams flow only in wet weather.

Sinkholes

Although caves and springs in karst areas may form scenic and otherwise desirable features, the sinkholes that sometimes accompany them
are often troublesome and may cause serious problems. A sinkhole forms
when the roof of an underground cavity becomes too thin and weak to
support the overlying beds and collapses (Fig. 5-8). Frequently this happens without warning, and there may be costly damage to overlying
buildings or other structures. Sometimes utility pipelines or cables may
be exposed or broken. Even in rural areas, the formation of numerous
sinks destroys the usefulness of the land. Such openings also provide
direct access to ground water and become avenues of pollution and contamination. Unfortunately some have been used as trash dumps, the
idea being to "get rid of the stuff and get the hole filled." In many places
the consequences have been extremely serious.

Although progress is being made in the use of geophysical and electronic methods, geologists have not yet perfected techniques for predicting the precise location or timing of sinkhole collapses. Until they do, we must learn to avoid building in vulnerable karst areas. We also must find ways to avoid the pollution and contamination of the state's many springs by better understanding their recharge areas and the rates of the underground water flow. Here indeed is reason for a growing field of geology called *hydrogeology* (literally the study of water moving through rocks).

Under certain conditions collapses may be triggered by changes in the position of the water table or by a change in the hydrostatic pressure of an artesian aquifer. In other words, the upward pressure of the water, or the buoyancy in the water table, may support the overlying beds until a change of some sort takes place. Many sinkholes have been observed to happen after extended periods of low rainfall. Sinkhole collapses also may be triggered by excessive rainfall.

In many areas construction and industrial activities have unintentionally stimulated or aggravated the formation of sinkholes. For example, streets, highways, and even airport runways have been built over undetected potential sinks that later collapsed under the weight and vibrations of traffic. In other cases the vibration and pounding from well-drilling activities have caused previously weakened spots to collapse. There are records of drilling rigs falling into these openings.[8]

Of all the geologic features in Missouri, those of karst origin are the most surrounded by mystery, awe, and superstition. Springs, caves, and sinkholes are often misunderstood by the uninformed. Even the early Greek writers of 300 to 500 B.C. were awed by the great volumes of water flowing continuously from the mouths of many caves. These writers believed that a vast body of water existed deep inside the earth from which the flow came and to which it returned by way of a connection from the floor of the ocean.

Several years ago a section of Kentucky Avenue in Columbia dropped, to leave a gaping hole about forty feet in diameter and some fifteen to twenty feet deep. It was immediately described by the press and other media as a "mystery hole" and attracted many visitors. It was, in reality, nothing more than another sinkhole like many others in the Columbia area and was very similar to the sinkholes in Rock Bridge State Park, just south of Columbia (Fig. 5–9).

Also among the "mysteries" are those broad, shallow topographic depressions with no apparent drainage and those large, deep chasms that sometimes open and swallow up a house, a section of the highway,

Fig. 5-9. Sinkholes in Boone County, near Rock Bridge State Park. The road intersection is at Pierpont.

or even a piece of an airport runway. These are not mysteries. They are the result of the karst processes just described. Missourians should be especially concerned with these phenomena because of the enormous area of the state underlain by the Ordovician and Mississippian carbonate rocks that allow Missouri to be "The Cave State" and to have so many large springs and sinkholes. The same processes that create the fantastic caverns and recreational waterways also make many areas in Missouri and in other states unsuitable for development of cities, streets, airports, and industrial parks.

Other Underground Spaces

Many underground openings and spaces are called caves, but some are really underground mines where limestone, dolomite, and sandstone have been removed (Fig. 5-10). Imaginative geologists, engineers, and architects have designed ways to use underground spaces by converting them to air-conditioned offices, warehouses, cold-storage facilities, or heated rooms, even manufacturing plants. These spaces offer distinct advantages: easily controlled, uniform temperature and humidity; freedom from surface noise; protection from the weather; lessened

Fig. 5-10. An underground limestone mine in Kansas City being prepared for further use as office space or storage. DNR–Vineyard.

risk of fire; easily maintained security; and solid, level floors and high ceilings. It is somewhat surprising at first visit to one of these spaces to find railroad boxcars and "eighteen wheelers" being loaded and unloaded many feet underground.

One of the pioneers in research on the uses of such space was Truman Stauffer, a geologist at the University of Missouri–Kansas City. John W. Whitfield (1981) surveyed the use of underground spaces in seven areas of Missouri. He evaluated ninety-two underground sites, of which sixteen had been improved and developed. He estimated the utilized space in these sites to be about 740 acres, or 1.2 square miles.

In the Kansas City area he described ten sites ranging from eight to more than three hundred acres, with ceilings ranging from twelve to eighteen feet in height. Roof thicknesses varied from fifty to one hundred feet. Room dimensions averaged about thirty to forty feet in length and width. These sites were being used as warehouses, office suites, cold-storage rooms, and a factory site for the manufacture of delicate scientific instruments. In the Kansas City area, the rock mined was the Bethany Falls Limestone of the Swope Formation, of Pennsylvanian age.

There are three underground sites in southwestern Missouri. One is in Springfield, where twenty acres of underground space were converted to a general warehouse with twenty-five-foot ceilings. Other space offers both refrigerated storage and heated rooms. At Carthage, thirty to forty acres include a general warehouse in addition to office suites and a large tennis court. A third site at Neosho contains about

Fig. 5-11. The view from the office facilities of Southwest Lime Company's underground warehouse in Neosho. The pillar in the office is Burlington-Keokuk limestone. DNR–James E. Vandike.

thirty to forty acres of office and warehouse space (Fig. 5-11). At Springfield and Neosho the rock removed was from the Burlington-Keokuk Formation. At Carthage it was the Warsaw Formation. The Carthage facility also has 5,000 square feet kept at 38 degrees F and used for beer storage, and a smaller room kept at minus 10 degrees F for frozen-food storage.

In the east-central area there are thirteen underground quarries, four in limestone, one in dolomite, and eight in St. Peter Sandstone. Only two, both in St. Peter Sandstone, are in use. One of these, at Crystal City, is a warehouse used for the storage of automobile tires. It has three tunnel-like spaces totaling nearly four acres with ceiling heights of twenty-five to forty feet. The other mine is a large tunnel about three hundred feet long and forty feet wide, with a ceiling height varying from fifteen to thirty feet. It has been a roller-skating rink for over twenty years.

Bussen Quarry in St. Louis County has developed a large block of rentable space that is easily accessible by rail, river, and highway. Also a development in Warren County will offer 250,000 square feet for refrigerated goods. As competition for surface space intensifies, such underground spaces are likely to become more valuable.

The use of underground space by man is not new. Robert F. Legget (1973) has pointed out that from the very earliest days of recorded history man has used "holes in the ground" for living, refuge, and burial purposes. These caves were mostly in limestone country, as are the caves and other underground spaces in Missouri. Legget cites the city of Petra in the desert of Jordan as a very early use of rock-cave dwellings and notes that the valley of the Nile, the "cradle of civilization," contains many temples and large rooms carved in solid rock. In particular, he refers to the two thousand years of protection provided for the Dead Sea Scrolls until their discovery in 1947 in caves near Qumran in Jordan by an Arab shepherd. Obviously the uniform temperature and humidity were factors in their preservation.

We are all aware that, as industry spread in our modern era, the use of underground space became more common with the development of tunnels for highways, railways, and buried utility lines. Yet it is still surprising to many that the Brunson Instrument Company planned underground space for the manufacture of delicate instruments in Kansas City as early as 1938, and the J. C. Nichols Company developed an underground industrial park using vacant limestone mines soon after.

6

THE MISSOURI COLUMN

The rocks of Missouri range widely in age, from the oldest Precambrian igneous rocks of the St. Francois Mountains, which are at least 2.5 billion years old, to modern sediments being deposited today in our streams and lakes. In this chapter we will take a closer look at the origins and characteristics of the rock formations found in Missouri.

In reading geological history, geologists move chronologically from the oldest rocks to the youngest. As mentioned earlier, the largest time divisions are eras, within eras are periods, and within periods are series. Each section of this chapter is divided and subdivided according to the time designations explained in Chapter 2. The Geologic Time Chart (Fig. 2-5) is an easy reference for these time designations. The complete nomenclature system is very complicated,[1] but we will stay within the following abbreviated plan, which is listed in order from the most general to the most specific.

Rock Term	Time Term
———	Era
System	Period
Series/Group	Epoch
Formation	———
Member	———

The geologic map of the state (Plate 3) gives an overall view of the areal extent of the rocks from each period. To look at the geology of a specific area, geologists use vertical representations, or columns, of the various strata of rocks and sediments, beginning with the oldest at the bottom and moving upward to the most recent deposits.[2] The standard symbols used to show these rock types are shown in Figure 6–1. Throughout this chapter, drawings using these symbols will represent the rocks as they can be seen in hillsides, road cuts, quarry faces, or even drill cores for wells. In some cases the drawings represent geologists' interpretations of both visual and other types of observations. A diagram of all the rock strata in the state would be many feet tall, but we have provided details from each of the periods in the sections where they are discussed.

In order to find information on a specific area, the reader should consult the geologic map (Plate 3) to determine the age of rocks in the area in question. Within the narrative descriptions for that period, it should then be possible to locate references to specific geographic areas. It is important to use the descriptive texts, the geologic map, and the diagrams of the columns in close conjunction with each other in order to obtain a full picture of the kinds of geologic features that exist throughout the state.

Precambrian Rocks

Precambrian rocks underlie all of Missouri but are exposed only in the southeastern area in the St. Francois Mountains and a few isolated localities nearby. They have been reached elsewhere in the state by deep drilling. They are not the oldest rocks in North America. As near as we can tell, the time we call Precambrian lasted more than 3 billion years. The early part of this time is called *Archaean* and the later part *Proterozoic*. The Missouri rocks are Proterozoic and are about 2 billion years old. In the northern and western areas, these rocks are known at depths ranging from 1,500 to 3,000 feet. In the Pea Ridge area of Washington County, they are at depths of 1,400 to 3,000 feet. Under St. Louis they are 4,000 feet below the surface. Richard F. Marvin (1988) determined the ages of Precambrian rocks in a few areas of the state, including Butler Hill Granite in St. Francois County at 1.5 billion years, Royal Gorge Rhyolite in Iron County at 1.53 billion years, a granite from Gentry County at 1.637 billion years, and a tuff from Osage County at 1.644 billion years.

Precambrian rocks crop out in the St. Francois Mountains over 5,000 square miles of the most rugged and inaccessible part of the state. In addition to the difficulty of access, the complex interrelationships of the different rock types have impeded and delayed a full and complete

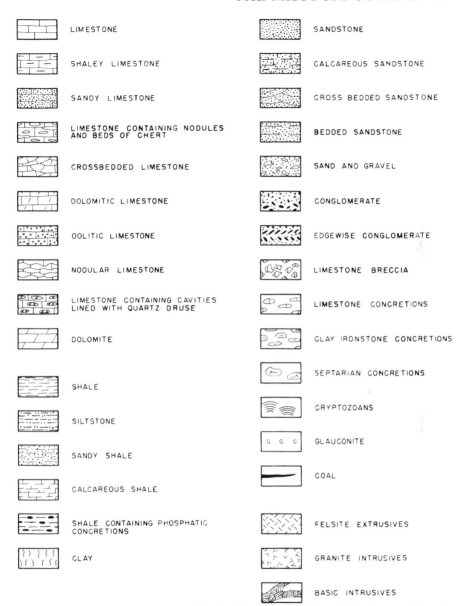

LIMESTONE

SHALEY LIMESTONE

SANDY LIMESTONE

LIMESTONE CONTAINING NODULES AND BEDS OF CHERT

CROSSBEDDED LIMESTONE

DOLOMITIC LIMESTONE

OOLITIC LIMESTONE

NODULAR LIMESTONE

LIMESTONE CONTAINING CAVITIES LINED WITH QUARTZ DRUSE

DOLOMITE

SHALE

SILTSTONE

SANDY SHALE

CALCAREOUS SHALE

SHALE CONTAINING PHOSPHATIC CONCRETIONS

CLAY

SANDSTONE

CALCAREOUS SANDSTONE

CROSS BEDDED SANDSTONE

BEDDED SANDSTONE

SAND AND GRAVEL

CONGLOMERATE

EDGEWISE CONGLOMERATE

LIMESTONE BRECCIA

LIMESTONE CONCRETIONS

CLAY IRONSTONE CONCRETIONS

SEPTARIAN CONCRETIONS

CRYPTOZOANS

GLAUCONITE

COAL

FELSITE EXTRUSIVES

GRANITE INTRUSIVES

BASIC INTRUSIVES

Fig. 6-1. Symbols commonly used by the Missouri Division of Geology and Land Survey to represent various rock types. DNR.

delineation and understanding of the Precambrian mass. Detailed studies have been made in only about 350 square miles.[3]

The oldest of the Precambrian rocks form a thick sequence of felsitic lava flows that were extruded over a surface that is not easily accessible to study. (*Felsite* is a term used to describe light-colored glassy to fine-textured igneous rock.) They are best exposed now between Ironton and Iron Mountain. Portions of these flows are very glassy, hard, and brittle and are called *rhyolite*. Most of them vary in color from light gray through pink to dark purplish red. In some masses there are scattered coarse crystals of light-colored feldspar. The crystals are called *phenocrysts*, and the rock is called *porphyry* (see Fig. 1–2). These early lava flows were subjected to a long period of erosion and were then covered by another felsitic flow that included some volcanic material from a vent in the area west of Lake Killarney. Then came another long period of folding and deformation.

After this period of deformation, or perhaps simultaneous with it, the area was intruded by a massive movement of deep-seated magma that cooled at depth, allowing the development of the coarse-grained rock called *granite*. Because of variations among the different parts of the magma, the result was a mass of granites of varying colors, some light to dark gray, some pink to red. The common minerals in the granites are quartz, feldspar, and mica. One of the best-known granites is the red to pink mass that is quarried at Graniteville and that forms the Elephant Rocks in the state park by that name in Iron County (see Fig. 3–2). Such large masses of granite are commonly referred to as *batholiths*.

Further deformation, folding, and uplift followed the intrusion of the granite batholith, and the older rocks were deformed to develop fissures, cracks, and crevices. Into these openings moved more molten magma, some of which reached the surface as flows, and some of which filled the crevices and cooled there. The latter formed dark-colored masses, some nearly vertical, which we call *dikes*, and some horizontal, which are called *sills*. These masses have a lower silica content than the granites (and thus are referred to as more *basic*) and are called *gabbro* or *diabase*. Following this period of intrusion and deformation, the entire region was uplifted and then uncovered by erosion until the coming of the Lamotte Sea, in Late Cambrian time, about 525 million years ago.

In addition to Elephant Rocks, the Precambrian igneous masses form other topographic and scenic features in Missouri. Among the best known are Taum Sauk Mountain, Mina Sauk Falls, Devil's Honeycomb, and the Devil's Toll Gate.[4] Another picturesque phenomenon that occurs among the igneous masses in the St. Francois Mountains is called a *shut-in*. Since igneous rock is more resistant to erosion than the overlying sedimentary beds, a stream or river can narrow, sometimes drastically, when it reaches the igneous rock. As large quantities of water from the wider streambeds are forced through these narrower channels, rapids are formed. These can greatly restrict the navigability of such streams but produce dramatic effects, especially as the water flows over and around igneous boulders in the midst of the channels. Thomas R. Beveridge (1978) reports forty-seven shut-ins in the Ozark region, of which Johnson Shut-Ins in Reynolds County is probably the best known (Fig. 6–2).[5]

The Precambrian igneous rocks, especially the felsites, are important to Missouri as a source of iron ore. Iron-rich minerals—magnetite and magnetic hematite—occur as vein fillings and as tabular masses. They have been found mainly in Iron Mountain and at Pilot Knob, with a few minor occurrences in the St. Francois Mountains. One large body of magnetite is mined 1,300 feet below the surface at Pea Ridge in Washington County. Other deeply buried magnetic masses are known in Crawford County at depths ranging between 1,700 and 3,500 feet. In addition, magnetic studies of Missouri have shown numerous other small areas, widely dispersed across the state, where the magnetic field is anomalously high, suggesting the presence of deeply buried magnetic deposits.

Paleozoic Era

The Paleozoic Era covered an enormous span of time—more than 320 million years—and is divided into the Cambrian, Ordovician, Silurian, Devonian, Carboniferous (including the Mississippian and Pennsylvanian), and Permian periods. During this time the North American continent was a part of Gondwana (see Fig. 2–1). In what is now the Midcontinent region, the earth's crust flexed, causing seas to migrate and many areas to be alternately submerged and drained. These shifting movements and marine migrations are recorded in the accumulated rocks.

By far the greatest part of the Paleozoic record in Missouri consists of carbonate rocks (limestone and dolomite) with interlayered shales and sandstones. This pre-Pennsylvanian column totals approximately 8,000 feet in aggregate thickness, but the total thickness is not present in any one location. The later part of the era (some 75 million years), the Pennsylvanian Period, was a time of widespread coal swamps and rapidly

Fig. 6–2. An aerial view of Johnson Shut-Ins in Reynolds County shows the narrowing of the East Fork of the Black River as it passes from the softer sedimentary rocks in the upper part of the photograph and reaches igneous rocks, which are much more resistant to the eroding force of the water than the sedimentary beds. U.S. Geological Survey.

migrating shallow seas. Only meager deposition occurred in Missouri during the late Pennsylvanian time of the Paleozoic Era, but it is well represented to the west and southwest.

Cambrian Period (Fig. 6–3)

For an enormously long span of time (perhaps 300 million years), the area that is now Missouri existed as a rugged region of barren granite and rhyolite knobs and hills with deep intervening valleys. It was barren because land plants had not yet come into being. In some places the local relief was as much as 1,500 feet (*relief* is the vertical distance between hilltop and valley floor). The evidence shows that there was weathering and erosion from Late Precambrian time until after the middle of the Cambrian. As the seas encroached upon this rugged surface in Late Cambrian time, the first sediment to be deposited was a sandstone conglomerate composed mostly of quartz grains with a mixture of larger fragments of quartz and feldspar and some fragments of the glassy igneous rocks. Obviously much of this conglomerate was derived from the weathering and erosion of the Precambrian granite, with its great abundance of quartz. The sandstone in Missouri is mostly gray to white in color, but in some areas it is yellow to red, a result of the weathering of iron-rich minerals.

This sandstone formation is from 200 to 300 feet in thickness but in a few places reaches nearly 500 feet. It is called the Lamotte Formation from its occurrence at the old Mine La Motte in Madison County. It is exposed only in the St. Francois Mountain area but extends outward to underlie younger rocks in most of Missouri. The few exceptions represent areas where the granite knobs were high enough to escape inundation by the Lamotte sea. Where the Lamotte Sandstone is exposed, it is quarried and used for building stone (see Fig. 2–3).

As the Ozark area continued to subside, the seas encroached farther upon the rugged topography, and the ruggedness began to be moderated by erosion. Consequently, the material being deposited in the sea became less well supplied with clastic sediment. Then Missouri became part of a shallow marine shelf that extended from New York to Mexico. Limestone and other carbonate materials began to accumulate. Many minor oscillations in the ocean depth led to the development of a thick sequence of carbonate beds with interlayered sandy and shaley beds. Some of the carbonate beds contain marine fossils.

The next well-defined formation above the Lamotte Sandstone consists of gray limestone and dolomite. This rock is mostly fine grained and

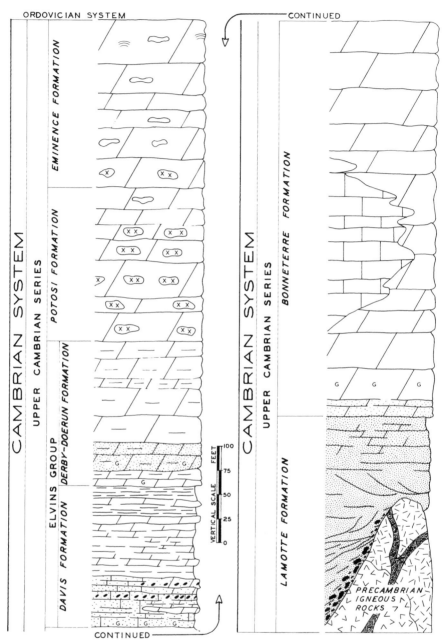

Fig. 6-3. Cambrian System, Upper Cambrian Series. DNR.

shaley but has some coarse zones. In areas where the igneous knobs or peaks were high enough to be out of the water during the earlier sea encroachment, these later beds lie directly upon the igneous rocks and contain sandy layers with some coarse fragments of the igneous rocks.

This formation is known as the Bonneterre. It is exposed widely in southeastern Missouri, mostly north and east of the St. Francois Mountains, and extends outward beneath most of the rest of the state. In the exposed area it is about 400 feet thick but has been found in drill holes to be over 1,500 feet.

The limestone and dolomite apparently represent relatively warm seas where many marine organisms lived. In some localities, fossil trilobites, brachiopods, and a few snails have been found. These are clearly Late Cambrian in age and are much like those known from similar beds in the Upper Mississippi Valley, in Tennessee, and in parts of New York. The Bonneterre Formation has been the source of most of the lead produced in Missouri.

For a long time after the deposition of the Bonneterre, the Ozark area remained submerged. Nonetheless, fluctuations in the sea level led to the accumulation upon the Bonneterre of about 200 feet of alternating thin limestones and shales. These layers, called the Davis Formation, have characteristics that suggest slow accumulation. One of these is the presence of a green mineral called *glauconite*, composed of iron and silica. The other is the presence of some beds composed of flat, disklike pebbles of limestone with rounded edges. These pebbles may have originated as polygonlike bits of limy mud formed among the cracks of mud flats exposed to temporary drying. The Davis Formation contains a fossil brachiopod known as *Eoorthis*, which is widely known in Late Cambrian beds elsewhere in the United States.

The Davis Formation is overlain conformably by a formation composed of dolomite layers that total about 150 feet. The lower of these layers are sandy; they then grade upward into thin-to-medium-bedded dolomite interbedded with siltstones and shale. These beds are mostly tan to brown in color. Near the top, one of the beds contains sponge spicules and fragments of echinoderms. This formation was once considered to be two separate formations; it is now recognized as one but still bears the two earlier names—hence the title *Derby–Doe Run*. Although most formations are named for a locality, these names derive from two mining companies that operated in southeast Missouri in the early 1900s.

Slowly, the fluctuation decreased and more stability returned, resulting in a massively thick-bedded carbonate called the Potosi Formation.

The Potosi Formation is variable in thickness, ranging from 70 to 300 feet, and it contains an abundance of small quartz crystals among red clay. In several localities, this formation has been the source of Missouri's barite.

The topmost part of the Cambrian rocks is a sandy dolomite that reaches a thickness of 350 feet and is called the Eminence Formation. It is abundantly supplied with chert, which may suggest a clearing and warming of the sea. The existence of sand suggests the possible uprising of an area to the west or north, which exposed older, clean sandstones. Many of the boulderlike masses of chert seem to have been formed as algal masses. The Eminence Formation is the host rock for many of the caves and springs that add so much to Missouri's popularity as a place for camping and other recreation.

The evidence seems clear that the close of the Cambrian saw a withdrawal of the seas that continued into early Ordovician time. Some parts of the latest Cambrian and some of the earliest Ordovician known elsewhere in the Midcontinent region are not represented in Missouri. It should be pointed out here that the thick masses of Cambrian dolomite were originally deposited as limestone but were altered over the millions of years of the Paleozoic to dolomite. During these same years, they were penetrated by mineralizing solutions that brought in the lead, zinc, copper, barite, and other elements that formed highly important ores.

Ordovician Period

For the several million years after the Cambrian, conditions in the area now known as Missouri changed very little. The seas remained mostly shallow, and the sea floor continued to subside and to receive large masses of carbonate sediment. There were, from time to time, temporary shallowings and even periods of slight exposure and erosion that furnished material for the deposition of sandy beds. These sandstone beds may represent the presence of migrating sea margins with accompanying beach conditions.

Gunter and Roubidoux sandstones are examples of those deposits. The result is that the Ordovician System consists of about 2,000 feet of cherty dolomites with interbedded sandstones and limestones (Fig. 6-4). This 2,000-foot thickness has been divided into eighteen formations that are grouped into three series, which are, in ascending order, the Canadian, Champlainian, and Cincinnatian.

Canadian Series (Fig. 6–5). For a brief part of early Ordovician time, the area that is now Missouri was not submerged. When the seas re-

Fig. 6–4. A typical exposure of Ordovician St. Peter Sandstone with overlying Joachim Dolomite, near Pacific.

turned, the initial deposit was a relatively thin, twenty-five-foot-thick sandstone called the Gunter. It grades upward and laterally into dolomite. As the seas deepened and stability was restored with slow subsidence, a few hundred feet of chert-rich limestone, called the Gasconade Formation, spread over the Ozark region. Minor shallowings occurred, and two sandstones interlayered with carbonate were spread over the area. The source was probably a rising of the Wisconsin dome. The resulting formation, now called the Roubidoux, is significant because, where they can be quarried, the sandstones are useful as building blocks. Many slabs are used as sandstone veneer over wood-frame houses. The bedding surfaces are strongly ripple-marked, indicating a shallow-water origin. Where reached by deeper drilling, the sandstones yield significant supplies of ground water.

The area continued to subside and remained as a marine shelf for a long time during which carbonate sediment continued to accumulate with only minor interlayering of thin sandstones and shales, along with large quantities of chert. This span of time saw the deposition above the Roubidoux (in ascending order) of the Jefferson City, Cotter, Powell,

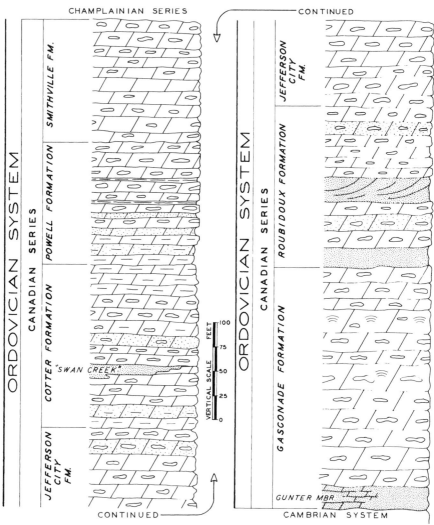

Fig. 6–5. Ordovician System, Canadian Series. DNR.

and Smithville formations. Collectively, from the base of the Gunter to the top of the Smithville, they constitute the Canadian Series, so named because of the similarity of this sequence to beds of corresponding age in eastern Canada.

Champlainian Series (Fig. 6–6). The seas were again drained from this area, but not for long. The earliest years of Champlainian time saw the deposition of thin, sandy carbonates, but conditions changed markedly,

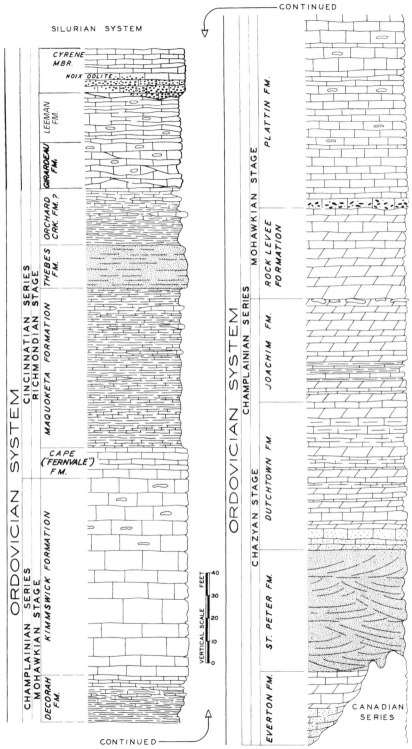

Fig. 6-6. Ordovician System, Champlainian and Cincinnatian series. DNR.

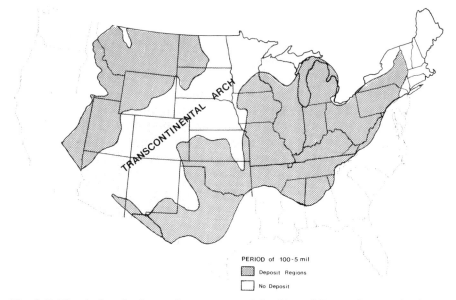

Fig. 6–7. The darker shading indicates areas of the United States that received sediments from late Precambrian to early Ordovician (Sauk sequence). Simplified from Sloss, 1988.

and a widespread erosion surface formed. This surface was then covered by a broad sheet of sand that spread widely over time and represents shallow, clear water near shore. This sand, known as St. Peter, is a distinctive formation of fine-to-medium-size, well-rounded quartz grains with frosted surfaces. It bears many features that suggest that the grains passed through a wind-blown environment. The formation is known to extend northward from Missouri to Minnesota, eastward into Illinois, and westward into Nebraska and South Dakota. In the western areas, it is somewhat shaley. Here it may have been deposited along the margin of the Transcontinental Arch, a broad area of the western continental region that was not submerged during the Ordovician (Fig. 6–7). In eastern Missouri, in the area of Festus and Crystal City, the formation is a clean and pure quartz sandstone that is 99 percent silica. Here it has long been mined and used for glassmaking, as a filter and molding sand, and for making silicate abrasives. Westward from there in Missouri, the formation thins, and in southern Boone County it exists only as isolated remnants.

After the St. Peter deposition, the seas became less clear and alternate deposition of carbonates and shale continued for a long time, resulting in the accumulation of formations now known as the Dutchtown, Joachim, Rock Levee, and Plattin. The Rock Levee is confined to

Cape Girardeau and Scott counties. All these formations are variable in thickness, and all carry marine fossils. One, the Dutchtown, is characterized by small cavities, or *vugs*, containing asphalt. The Plattin differs somewhat from the others in that it is a relatively pure limestone and its fossils seem to represent a return to a shallow marine environment.

There was a distinct change in the marine environment over much of the Midcontinent area after the deposition of the Plattin. Apparently the Transcontinental Arch became a source of sediment and the Plattin and earlier formations in Missouri became covered with some forty or more feet of thin-bedded, silty, clay-rich fossiliferous limestone. This formation is called the Decorah because it appears to be a southward extension of a limestone and shale sequence previously described near Decorah, Iowa. Along with other beds of similar age in the Midcontinent region, this formation contains thin layers of bentonite, an altered volcanic ash believed to have originated from the contemporaneous orogenic activity (the folding, thrusting, and faulting processes that created mountains) in the Taconic Range of eastern New York. The uppermost layers of the Decorah are relatively pure limestone, indicating that the seas were clearing. These layers continue upward into the overlying Kimmswick Formation, which is an unusually pure limestone consisting of 95 to 99 percent pure calcium carbonate.

The Kimmswick is very fossiliferous; among its fossils is a distinct form commonly called the Sunflower Coral because of its disklike shape and distinctive markings (Fig. 6-8). Ranging from a few inches to more than a foot in diameter, Sunflower Coral is in reality a fossil marine colonial algal mass, which thrived in a warm sea. Until recently, this form was called *Receptaculites*, but it is now known as *Fisherites*.

Cincinnatian Series (see Fig. 6-6). There is strong evidence for the beginning of an upwarping of the Ozark region as the Champlainian came to a close. The top of the Kimmswick is an erosion surface upon which there is a shaley, thin-bedded limestone of limited area called the Cape. This upwarping limited the later Ordovician deposition in Missouri to only the easternmost areas but allowed it to spread widely across other parts of the Midcontinent. In Missouri there was a return to muddy, silty deposition that formed a thick mass of dull green to gray calcareous shale now called the Maquoketa Formation, which thickens northward into Iowa. This shale contains distinctive fossils called *graptolites*. These are common index fossils in Ordovician and Silurian beds worldwide. On bedding planes in gray or black shales, they look like pencil markings, or in some forms they resemble broken bits of scroll-saw blades (Fig. 6-9).

Fig. 6–8. *Fisherites* from the Kimmswick Formation. Commonly called Sunflower Coral, *Fisherites* is most likely a colony of marine algae.

As the upwarping continued, some localized areas received abundant supplies of sand. Above the Maquoketa, and interfingered with its upper layers, is a sandstone called the Thebes. As the sand supply was exhausted, mud and silt returned to become the dominant material, and a bluish gray shale called the Orchard Creek was formed. Its age has been debated for some time.

Silurian Period

Except for the questionable Orchard Creek Formation, mentioned above, the whole of Missouri and much of the Midcontinent appears to have been exposed to a considerable period of uplift and erosion, probably for as long as 80 million years. The Silurian and Devonian periods, which are widely represented elsewhere in the United States by thick and extensive limestones, are meagerly represented here.

Silurian rocks appear at or near the surface in only two widely separated areas in Missouri. One is in the southeastern counties of Cape Girardeau and Perry, from which the beds dip eastward under younger beds into Illinois. The southeastern Missouri beds can be roughly cor-

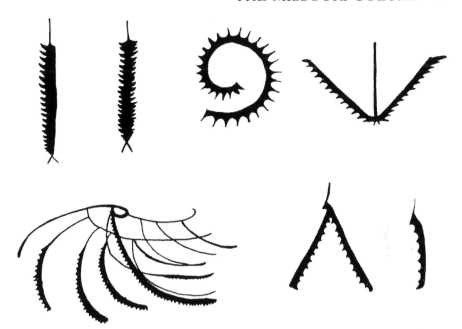

Fig. 6–9. Impressions of Ordovician graptolites as seen on bedding planes of gray shales. Natural size.

related with the Alexandrian and Niagaran series of the northern and eastern United States. In the southeastern counties, only two formations are currently recognized as being Silurian. The lower one is called the Sexton Creek. It is about thirty to sixty feet thick and consists of olive gray cherty limestone, which is irregularly and thinly bedded. It contains an abundance of chert in the lower part and some interbedding of green shale. The upper unit, known as the Bainbridge, is a dark red clayey limestone that ranges from 30 to 160 feet thick. Some light gray beds are mottled with purple and green. The green mineral called glauconite is common in the basal portion.

The other area of Silurian is in the northeastern counties of Lincoln, Pike, and Ralls, where there are scattered outcrops over the Lincoln fold. Here, two Silurian units are recognized. The lower is called the Bowling Green. It is a medium- to thick-bedded, yellowish-brown to tan, finely crystalline dolomite that ranges up to sixty feet in thickness. Where the thicker portions occur, this stone is quarried for agstone and "road metal," or crushed stone placed on roadways as a driving surface or to underlie asphalt or concrete paving. In some quarry faces, cavities a few inches in diameter are filled with petroliferous material that leaves a dark "oil stain" when it drips down the quarry face.

The upper bed consists of white to light gray, finely crystalline lime-stone with slabs of white chert. Exposures are scattered and not very good, and the thickness is estimated to be only about ten to fifteen feet. This limestone is correlated with the Sexton Creek Formation of the southeastern part of the state.

This limited distribution of Silurian rocks may be, to a large extent, due to upward movement of the Ozark region and/or to the rising of a broad area referred to as the northeast Missouri arch. Also, it is possible that Silurian deposits were removed during the multimillion-year pe-riod of erosion between Late Ordovician and Middle Devonian.

Over most of the Midcontinent area, the youngest Ordovician rocks and the oldest Silurian are separated by an unconformity that may rep-resent millions of years. Paleontologists recognize in North America a discontinuity between Late Ordovician and Early Silurian brachio-pods. P. M. Sheehan (1973) suggested that this unconformity may have occurred because the sea level had been lowered for a long time by a contemporaneous glacier when the continents were still united as Gondwana. Glacial deposits in Africa and South America that may cor-relate with this period provide additional evidence for his theory.

Devonian Period (Fig. 6–10)

In the Midcontinent region, and especially the area that is now Mis-souri, the Devonian Period was a time of shifting seas, changing configu-ration of the sea floor, and varying types of sedimentation. It was appa-rently a time of warm climate, as evidenced by the variety of marine fossils and ancient coral reefs. The coral *Hexagonaria* is common (Fig. 6–11). The Devonian rocks are almost entirely limestone and dolomite.[6]

As a result of the changing conditions, the Devonian beds are not continuous and cannot be followed or traced over wide areas. Instead they are characterized by many masses, which are thick in their central portions but have thin edges and a lithology that changes over relatively short distances. In all cases the bottom Devonian beds are unconform-able upon older beds, mostly on Ordovician, which indicates a long pe-riod of erosion or no deposition. As stated earlier, the Silurian is con-spicuous by its near absence.

The Devonian deposits of southeastern Missouri are limited to Ste. Genevieve, Cape Girardeau, and Scott counties. They comprise a thickness of 700 to 1,000 feet of limestone, but because they are com-plexly folded and faulted, their characteristics and stratigraphic succes-sion are difficult to determine. They have been divided into six forma-

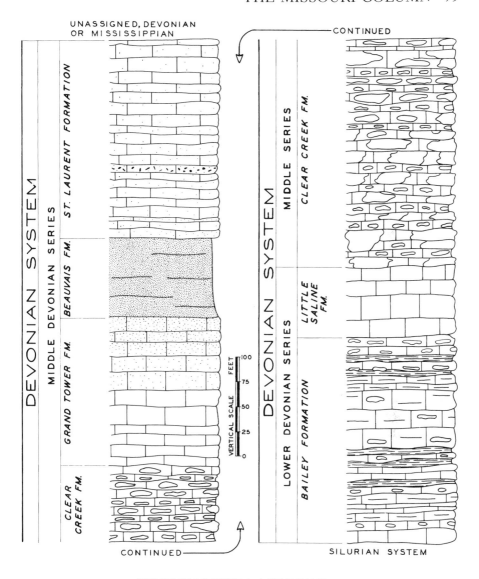

CONTINUED

DEVONIAN SYSTEM

MIDDLE DEVONIAN SERIES

ST. LAURENT FORMATION

BEAUVAIS FM.

GRAND TOWER FM.

CLEAR CREEK FM.

VERTICAL SCALE · FEET

100

75

50

25

0

CONTINUED

CONTINUED

DEVONIAN SYSTEM

MIDDLE SERIES

CLEAR CREEK FM.

LOWER DEVONIAN SERIES

LITTLE SALINE FM.

BAILEY FORMATION

SILURIAN SYSTEM

SOUTHEASTERN MISSOURI

Fig. 6-10. Devonian System, Lower and Middle Devonian Series. DNR.

tions called, in ascending order, the Bailey, Little Saline, Clear Creek, Grand Tower, Beauvais, and St. Laurent. All are limestones except the Beauvais, which is about eighty feet of white sandstone that appears to have been derived from erosion of the older St. Peter Sandstone. The Bailey and Clear Creek are cherty and fossiliferous. The Little Saline is

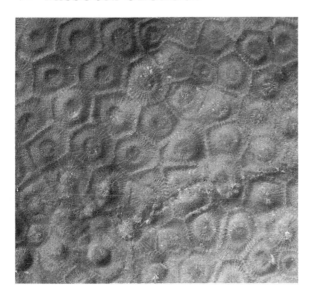

Fig. 6–11. The *Hexagonaria* coral common in the Callaway Limestone of Missouri.

limited to Ste. Genevieve County and has been quarried and marketed commercially as Rose Marble. The Grand Tower contains numerous small coral reefs and has been used architecturally as Golden Vein Marble. These Devonian beds cannot be traced to other parts of the state.

In central and east-central Missouri, the Devonian rocks are Middle to Late Devonian in age and are relatively thin compared to those in the southeastern section. Their outcrop area is essentially limited to a narrow belt north of the Missouri River between southern Boone County and Montgomery County, and south of the river in isolated spots in Pettis, Cooper, and Moniteau counties. The total thickness of the section rarely exceeds fifty feet.

The limestones in these counties have been variously interpreted but seem to be separate aspects of one formation. Formerly thought of as separate formations, the beds vary horizontally and vertically over short distances and interfinger so thoroughly that only one name is needed, and they are now called the Callaway Formation. The total sequence represents a clear, warm marine environment that changed in depth from time to time. The beds are fossiliferous and can be correlated with the Middle Devonian Cedar Valley Formation of Iowa.

In the Devonian rocks we find the first widespread black shales containing carbonized material, because land plants did not develop until Devonian times. Throughout much of the Midcontinent region, in the sequence of beds between those that are definitely Devonian and those that are definitely Mississippian, there are significant thicknesses of black to dark gray shales. Some of these shales can be traced eastward into the eastern highland areas, and the carbonized plant material in them may have derived from the Devonian forests, which came into being in the Catskill area of New York.

In Callaway and Montgomery counties the Callaway Formation is overlain by sixty feet of gray-green shale that in some places is sandy and in others calcareous with thin limestones. This is called the Snyder Creek Shale. The upper part is fossiliferous, and the fossils are definitely Upper Devonian in age.

In east-central Missouri, through Franklin, St. Charles, St. Louis, and Warren counties, and extending easterly into Jefferson County, there is a light to medium gray to yellowish gray oolitic limestone named the Glen Park Limestone. It can be traced into northern Ste. Genevieve County, where it is only fourteen inches thick and is probably an extension of the Snyder Creek.

The Grassy Creek Shale is dark olive to greenish black and is limited to Marion, Ralls, and Pike counties. In Pike County a fourteen-foot thickness of green to bluish gray sandy shale is called, by some writers, the Saverton. Above both of these shales, in a few localities, is a brittle blue-gray limestone called the Louisiana. It has a strongly developed joint system along which ground-water solution has created numerous mazelike caves. The most famous is the Mark Twain Cave near Hannibal where Tom Sawyer and Becky Thatcher got lost.

In southwestern Missouri, in Barry and McDonald counties, are exposures of a black, very thinly layered, spore-bearing, carbon-rich shale. The maximum thickness is about thirty feet. This shale is also known along the James River in Christian County and in wells in Greene and Christian counties. Thick intervals of similar shale have been found in Ohio, Indiana, Illinois, Tennessee, and Arkansas. That in Missouri is continuous with that in Arkansas, which in turn is traced into that in the vicinity of Chattanooga, and the name *Chattanooga Shale* has been adopted in Missouri. This shale rests on Ordovician dolomite and is overlain by early Mississippian beds, but its age has been debated for some time, leading some writers to refer to the "Black Shale Problem." That in Missouri is now regarded by the Missouri Geological Survey as being Devonian.

In Holt and Atchison counties in northwestern Missouri, there is an interval of 150 to 200 feet of black, very thinly layered, spore-bearing, carbonaceous shale. It thins to an edge in Clay, Caldwell, Daviess, and Harrison counties. Although there is no conclusive evidence for the selection of a name, this shale is provisionally called the Kinderhook

Shale. Some researchers have thought it to be related to the Boice Formation of Nebraska and Kansas. Elsewhere the term *Kinderhook* is used for early Mississippian beds.

Mississippian Period

Shifting seas, oscillating sea level, and instability of the sea floor at the close of Devonian time and the beginning of Mississippian resulted in a sequence of rock units that are complexly interlayered and that vary from one area to another. As a result, geologists have found it difficult to make precise correlations between regions. Some beds with definite characteristics in one locality are considerably different even a short distance away and thus cannot be reliably traced. Many of the beds are easily eroded and become covered without forming good exposures. Also some beds lack distinctive fossils, some have no fossils, and age determinations have been difficult. Because of these conditions, many geologists in Missouri have, in recent years, chosen to avoid making definite age assignments to these beds and instead recognize them as possible transition, or Devono-Mississippian, units. However, Thomas L. Thompson (1986), after a thorough review of the literature and extensive fieldwork, has presented a complete revision of the Mississippian succession with new definitions of some old units and introduction of some newly named ones (Table 6–1; Fig. 6–12).

Mississippian rocks have been of extreme importance to the state of Missouri because it is this system that contained the great deposits of lead and zinc ore that made the state part of one of the most productive mining districts of the world and for many years enabled Missouri to lead the United States in the production of zinc. Mississippian limestones are also quarried and used for agstone and concrete aggregate. In addition, Mississippian limestones from Greene and Jasper counties have been quarried for building stone and ornamental "marble" that have been used throughout the United States.

During Mississippian time the shifting seas and changing environments continued on a broader scale. The rocks of this age are at, or near, the surface in about one-fourth of the state. The formations vary both laterally and vertically, and their successions differ from one part of the state to another. Throughout most of the Midcontinent, the Mississippian is divided into four series—the Kinderhook, Osage, Meramec, and Chester—that appear to have been deposited between the rising Ozark dome and the Transcontinental Arch. Figures 6–13 through 6–16 present portions of the succession of Mississippian formations in

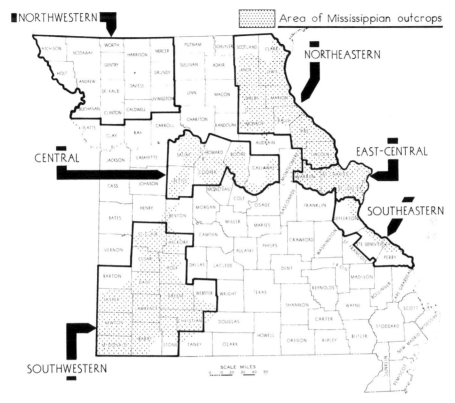

Fig. 6-12. The regional distribution of the Mississippian System in Missouri. DNR.

four parts of the state and indicate the extent to which the formations can vary from region to region.[7]

Kinderhook Series. The type area of the Kinderhook is in northeast Missouri and adjacent areas in Illinois and Iowa. Here it consists mostly of the Hannibal Formation, a clayey sandstone and fissile siltstone. The bedding planes of the Hannibal often have "Roostertail" (*Taonurus*) markings (Fig. 6-17) and irregular worm trails and "borings." Above the Hannibal is a sequence of limestones referred to as the Chouteau Group. From the northeastern region this group can be traced southwestward to the Ozark area, where it is made up mostly of interbedded limestones. The most dominant of these are the Compton, Sedalia, and Northview.

Osagean Series. In the Mississippian of Missouri the Kinderhook is followed by a thick limestone section comprising the Osagean Series.

TABLE 6-1

Formally recognized stratigraphic units of Mississippian age in Missouri, by geographic region.

	East-central	Southeastern	Northeastern
CHESTER		Vienna Limestone Tar Springs Sandstone Glen Dean Formation Hardinsburg Formation Golconda Formation 　Haney Limestone Mbr. 　Fraileys Shale Member 　Beech Creek Ls. Mbr. Cypress Formation Paint Creek Formation 　Ridenhower Ls. Mbr. 　Bethel Member 　Downeys Bluff Ls. Mbr. Yankeetown Sandstone Renault Formation	
MERAMEC	Ste. Genevieve Ls. St. Louis Limestone Salem Formation Warsaw Formation	Aux Vases Sandstone Ste. Genevieve Ls. St. Louis Limestone Salem Formation Warsaw Formation	Ste. Genevieve Ls. St. Louis Limestone Salem Formation Warsaw Formation
OSAGE	Keokuk Limestone Burlington Limestone 　"lower Burl. Ls." Fern Glen Formation 　Meppen Ls. Member	Keokuk Limestone Burlington Limestone 　"lower Burl. Ls." Fern Glen Formation 　Meppen Ls. Member	Keokuk Limestone Burlington Limestone
KINDERHOOK	 Bachelor Formation Bushberg Sandstone	Chouteau Group undiff. Bachelor Formation Bushberg Sandstone	Chouteau Group "McCraney Limestone" "Chouteau Limestone" Hannibal Shale Horton Creek Limestone

Adapted from Spreng, 1961.

Southwestern	Northwestern	Central
Fayetteville Formation (Wedington Ss. Mbr.) Batesville Formation Hindsville Limestone (Carterville Formation)		
	St. Louis Limestone Salem Formation	Salem Formation
Warsaw Formation	Warsaw Formation	Warsaw Formation
Keokuk Limestone Short Creek Oolite Mbr.	Keokuk Limestone Short Creek Oolite Mbr.	Keokuk Limestone
Burlington Limestone Elsey Formation Reeds Spring Formation	Burlington Limestone Elsey Formation	Burlington Limestone
Pierson Limestone	Pierson Limestone	"Pierson Limestone"
Chouteau Group Northview Formation Baird Mountains Ls.	Chouteau Group Northview Formation Sedalia Formation	Chouteau Group Northview Formation Sedalia Formation "Unnamed Limestone"
Compton Limestone Bachelor Formation	Compton Limestone Bachelor Formation	Compton Limestone Bachelor Formation

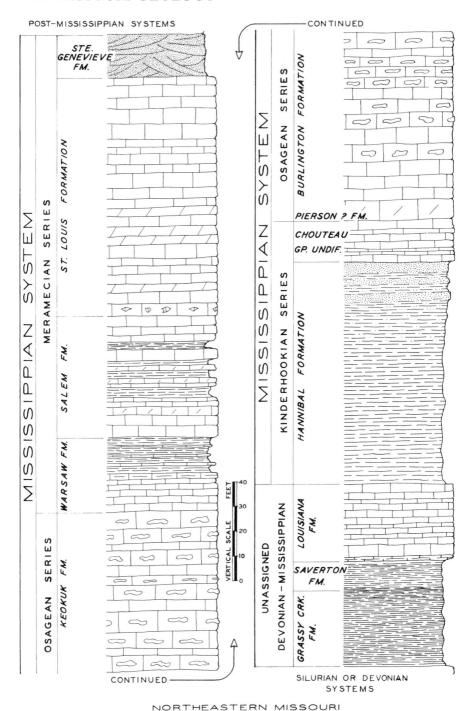

Fig. 6–13. Mississippian System, Kinderhookian, Osagean, and Meramecian series and unassigned Devonian-Mississippian formations in northeastern Missouri. DNR.

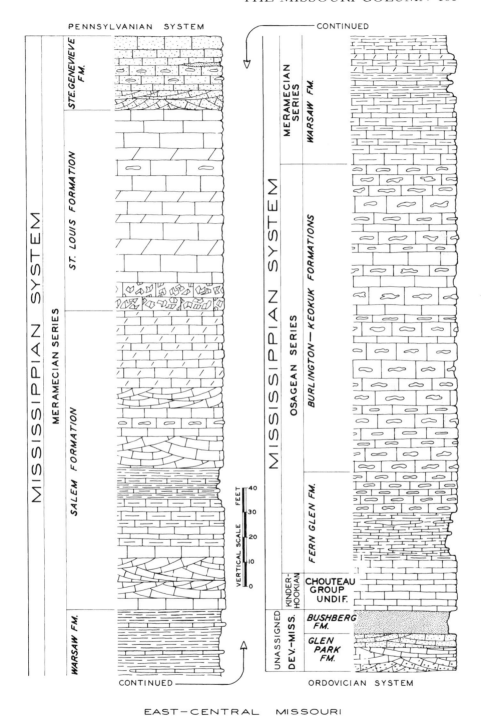

Fig. 6–14. Mississippian System, Osagean and Meramecian series and unassigned Devonian-Mississippian formations in east-central Missouri. DNR.

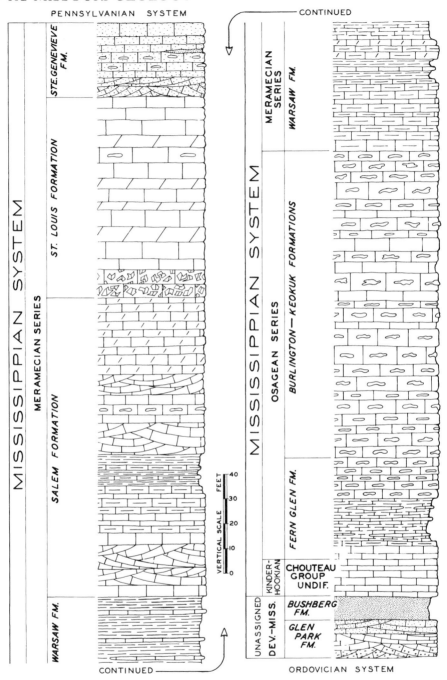

Fig. 6-15. Mississippian System, Kinderhookian, Osagean, Meramecian, and Chesterian series and unassigned Devonian-Mississippian formations in southwestern Missouri. DNR.

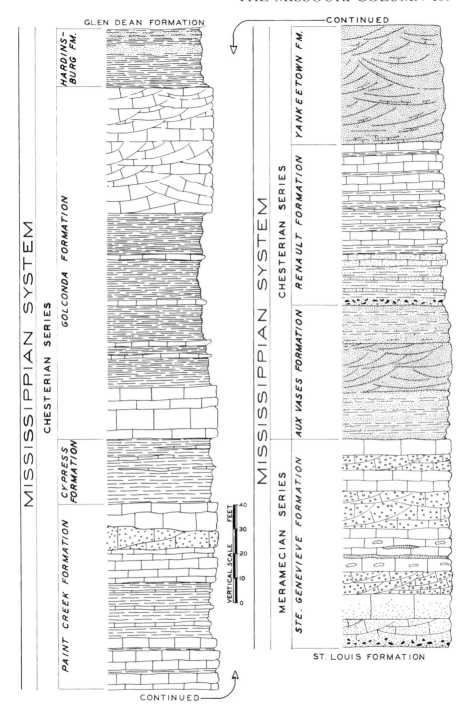

Fig. 6–16. Mississippian System, Meramecian and Chesterian series in southeastern Missouri. DNR.

Fig. 6–17. "Roostertail" markings are common on the bedding planes of siltstones in the Kinderhook beds.

This section is present as a broad band surrounding the older beds of the Ozark region. It varies laterally and vertically in lithology because the shifting of the seas caused local variations in the deposits. However, these can be reliably correlated and traced from southeastern Iowa into central and southwestern Missouri and on into Kansas and Arkansas. The most complete section of the Osagean is in the southwestern part of the state.

The Burlington Limestone is the most prominent formation. It was recognized in the mid–1800s and named for its occurrence in the bluffs along the Mississippi Valley at Burlington, Iowa. It is uniformly a crystalline, white to light brown, crinoidal limestone, and it contains an abundance of light gray to white chert (flint) in nodules of assorted sizes and in many thick layers (Plate 7; see also Fig. 3–5). When the chert is removed, the limestone is about 95 percent calcium carbonate. In addition to crinoid fragments it also contains brachiopods, corals, and bryozoa. Rarely, trilobites and sharks' teeth are found. Throughout northeastern Missouri and into the central part of the state, the Burlington ranges from seventy-five to one hundred feet in thickness.

Also getting its name from an Iowa location, the Keokuk is in many ways much like the Burlington, and the two are sometimes mapped together. The Keokuk is also coarsely crystalline, well bedded, light

Fig. 6–18. A specimen of *Archimedes*, or the fossil corkscrew, from the Warsaw Formation.

bluish gray, and abundantly supplied with chert. The chert is more common in the lower part. The fossils are similar, except that the Keokuk also bears the unusual bryozoa-algae combination fossil called *Archimedes*, or the fossil corkscrew (Fig. 6–18). This fossil can be seen in many places in the walls and floors of the Capitol in Jefferson City, where it occurs in the polished marble (Fig. 6–19). In the northeastern and central parts of the state, the Keokuk lies above the Burlington, but, as already stated, the point of contact is difficult to determine precisely. Both are about the same thickness, and where both are present the total thickness is more than one hundred feet. Toward the southwestern part of the state, the Burlington thins to nothing and the Keokuk rests upon the Grand Falls or the Reeds Spring beds. Other less significant beds are locally recognized in the Osage.

Meramecian Series. Except in the southeastern and southwestern areas of the state, the Osagean Series is followed conformably by the Meramecian. The lowermost formation in the Meramecian is the Warsaw, named for Warsaw, Illinois. This formation can be found over a wide area of Missouri, from east-central to southwest, but in the northeastern region it occurs only in scattered outcrops in Lewis and Clark counties. There it is about forty feet thick and is primarily a finely to coarsely crystalline limestone, with geodes being common in the lower part and the upper part consisting mostly of shaley limestone. From

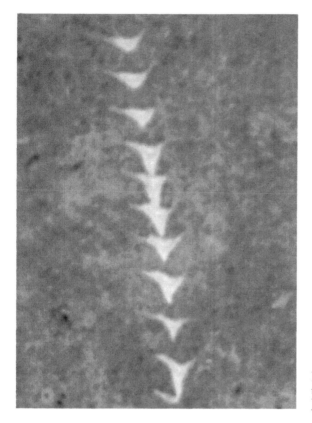

Fig. 6–19. *Archimedes* in the wall of the Capitol in Jefferson City.

this area it can be traced southwestward along the western flanks of the Lincoln fold to St. Charles and St. Louis counties.

Across the rest of the state the Warsaw Formation varies from a dark fissile shale with paper-thin layering through a coarsely crystalline limestone. In southwestern Missouri it is a medium gray, finely crystalline, thick-bedded limestone (Fig. 6–20). For many years the Warsaw was quarried, trimmed to dimensional blocks, cut, and polished for use as a building stone known as Carthage Marble. It also contains the fossil *Archimedes*.

Across most of the state, the Warsaw is followed by the Salem Formation, which is mostly limestone, partly dolomitic. In some areas it contains Syringoporoid corals. Where it is best developed, in the east-central area, it may be as much as one hundred feet thick. In most of the area, the Salem is overlain unconformably by the St. Louis Limestone, which is also best developed in the east-central region. Here some beds are lithographic. Characteristic fossils in the St. Louis Limestone are Lithostrotionellid corals (Fig. 6–21).

The uppermost Mississippian that can be recognized in northeastern

Fig. 6–20. Mississippian limestones overlying Devonian shale along Highway 59 near Noel, in extreme southwestern Missouri.

Fig. 6–21. A typical colony of Lithostrotionellid corals. These are common in the St. Louis Limestone of Missouri.

Missouri is the Ste. Genevieve, which here is thin, only one to four feet thick, and consists of white sandstone and very fine-grained limestone. Elsewhere, in east-central and southeastern areas of the state, it is a massively bedded white sandy limestone with some coarsely crystalline and some oolitic beds. Near the middle it commonly contains red and gray chert.

Chesterian Series. The shifting seas and oscillating sea level became more active in Chesterian time. These beds in southern Missouri are, as A. C. Spreng (1961) put it, "crudely rhythmic repetitions of sandstone, shale, and limestone." In southeastern Missouri they are limited to Ste. Genevieve and Perry counties, where they are exposed in bluffs along the Mississippi River. They are considerably faulted and dip steeply. Because of the deformation, the beds cannot be traced very far, and geologists must rely on comparisons with the less-disturbed section of these formations in Illinois. These beds form the thickest Chesterian section in the state and have been divided into the Aux Vases, Renault, Yankeetown, Paint Creek, Cypress, Golconda, and Hardinsburg formations.

In southwestern Missouri the Chesterian is strikingly different than in the southeast. Here it occurs mostly as isolated exposures and outliers of formations more widely developed and known in Oklahoma and Arkansas.

The Hindsville Formation is named for exposures in northern Arkansas. In Missouri it is found as discontinuous isolated remnants overlying the Keokuk. It is mostly an oolitic dark gray limestone interbedded with light gray calcareous siltstone. It contains fossils that denote an early Chesterian age. One of the best exposures is near Washburn in Barry County, where it is about sixty feet thick, but much of the section is covered. It can also be seen on the southwestern side of Oakleigh Mountain.

Above the Hindsville, and apparently transitional with it, is the Batesville Formation. It is also named for an Arkansas locality. It is mostly yellowish brown, evenly bedded calcareous sandstone with thin beds of gray oolitic limestone. The thickness is about thirty to fifty feet. In localities where the Hindsville is absent, the Batesville lies upon the Keokuk. It contains Chesterian brachiopods and pelecypods.

The Carterville Formation was named for isolated patches of varying lithologies, near Carterville, which do not fit the usual geologic sequence pattern. Similar unusual occurrences are known in filled sinks, prospecting pits, and in drill holes in Jasper, Newton, and western Lawrence counties. Thomas L. Thompson (1972 and 1986) interpreted these occurrences as isolated remnants of once more extensive beds that can be correlated with the Hindsville and Batesville, and perhaps even the Warsaw. Where they are fossiliferous, the fossils are clearly Chesterian.

Also exposed on Oakleigh Mountain in Barry County, and on Reed and Lennox mountains, is the Fayetteville Formation. It is mostly black fissile carbonaceous shale about twenty feet thick and is somewhat interbedded with dark gray to black limestone. On Oakleigh Mountain the Hindsville and Batesville are overlain by some 125 feet of brown, cross-bedded sandstone containing plant fossils. This sandstone has been interpreted as being part of the Wedington, originally described southeast of Fayetteville as a member of the Fayetteville Formation.

Pennsylvanian Period

Between the deposition of the Mississippian and Pennsylvanian rocks, widespread changes occurred in the structural character of the Midcontinent region. The region was mostly uplifted and exposed to erosion so that the earliest Pennsylvanian beds rest unconformably upon Mississippian and older rocks. Even after the beginning of Pennsylvanian deposition, this instability continued so that the shifting of the seas, rising and falling of the continents, and relatively rapid alternation of marine and nonmarine environments caused deposition of sequences of sediments with features unlike those from previous periods of the Paleozoic.

It was during this time that different parts of the continent rose and fell at different rates and many individual basins were formed. Sea-level

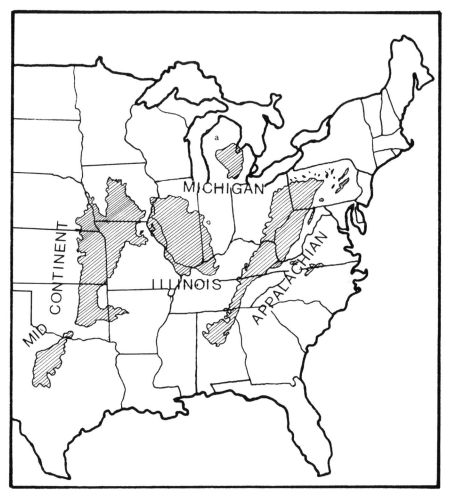

Fig. 6–22. Pennsylvanian coalfields of the eastern United States.

fluctuations may have been caused by alternating Gondwanian glacia-
tions. Contrary to earlier times, when the oceans simultaneously covered areas of several states, differential warping of the crust allowed for
many large areas to be invaded by the seas while intervening areas were
dry land or swamps.

During this time coal-producing swamps formed in many basins, from
the eastern United States as far west as Oklahoma and Texas and south
to what is now Alabama (Fig. 6–22). Some of these areas at times resembled today's Okefenokee and Great Cypress swamps of the southeastern United States or Big Oak Tree State Park in the Missouri Bootheel.

Because of fluctuations within localized areas, we find discontinuous layers of marine and nonmarine sediments ranging in area from a few counties to large portions of a state. Nevertheless, because of generally similar climatic conditions across the continent, it is possible to make time and age correlations from basin to basin. The rising and falling movements in the basins resulted in the deposition of sequences of sediments in which marine and nonmarine beds alternate and interfinger. For any one basin the rock record ideally contains repetitive sets such as those depicted in Figure 6–23.

These repetitive sets have been called *cyclothems* by many geologists. Many are distinctive enough to be recognized and mapped as formations. They may vary in thickness from a few feet to more than sixty. Many bear the name of a coal because the coals were named and mined before the nature of the sequence was well understood. The best development of the cyclic repetitions in Missouri is within the Desmoinesian Series.

Rocks of the Pennsylvanian contain the important coal beds of the region described above, and many important limestones, especially in western Missouri. In addition the Pennsylvanian has furnished some states, especially Missouri, with high-grade refractory clays used in making liners for blast furnaces, launching pads for rockets, boiler liners for steam-powered ships, and other industrial uses where high temperatures must be contained.

In Missouri the aggregate thickness of the Pennsylvanian rocks is about 2,000 feet, but there is no one place where this section is complete. As a result of rapid alternation of conditions, at least seventy-three formations may be recognized, and within some of these formations, several distinct lithologic units (called *members*) can be identified. For example, the Stanton Formation contains three limestones, one green sandy shale, and one black-to-gray phosphatic shale, all in a thickness of thirty-five feet. The Topeka Formation, also only thirty-five feet thick, contains nine members. While the Ordovician, which lasted approximately 70 million years, gave Missouri nineteen formations, Missouri's seventy-three Pennsylvanian formations developed in only 45 million years.

The Pennsylvanian beds can be found in more than 65 percent of Missouri counties surrounding the Ozark area, and the existence of isolated remnants in the Ozark area suggests that the Pennsylvanian seas may have at one time been continuous over the state except for a few isolated knobs of the St. Francois Mountains.

The Pennsylvanian rocks have been divided into five series, which

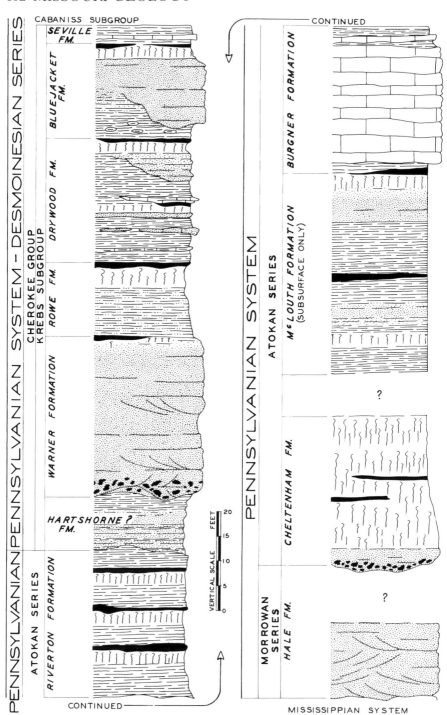

Fig. 6-23. Pennsylvanian System, Morrowan, Atokan, and Desmoinesian series (Cherokee group, Krebs subgroup). DNR.

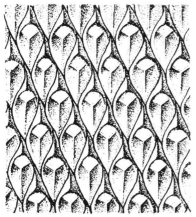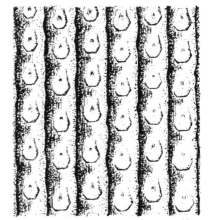

Fig. 6–24. Patterns left by scalelike leaves on the trunks of Pennsylvanian age trees. Left, *Lepidodendron*; right, *Sigillaria*.

are of unequal distribution and thickness. The oldest of these is the Morrowan, and it is followed in decreasing age by the Atokan, Desmoinesian, Missourian, and the Virgilian. In some areas these series are further subdivided into groups and subgroups; samples of such subdivisions are illustrated in Figures 6–23 and 6–25 through 6–28.[8]

Morrowan Series (Fig. 6–23). This series is represented in Missouri only by disconnected sandstone outliers (or isolated remnants) of the Hale Formation, which is well developed in northwestern Arkansas. These outliers are limited to Barry County, where they cap Oakleigh Mountain, Lennox Mountain, and Reed Mountain. They consist of a brown and yellowish brown, medium-grained, quartzose sandstone that contains casts and molds of plant fossils, including *Lepidodendron* and *Sigillaria* (Fig. 6–24). The maximum thickness is about sixty-five feet.

Atokan Series (see Fig. 6–23). The Pennsylvanian rocks assigned to Atokan time are among the most perplexing beds in the state. They are not continuously represented vertically or laterally in any area and vary considerably in character. Although five formations have been recognized and defined, there is some doubt that the assignments and correlations are correct.

Cheltenham, after an area near St. Louis, is the name applied to separate masses of clay with associated clastic sediments. The formation includes plastic clays, flint clays, and burley clays, all refractory, all varying in color from white to purplish red. All of these beds are economically important. They all rest upon rocks that are clearly older than earliest Pennsylvanian but do not contain fossils that permit close correlations with beds for which the age is better known.

In many basins, as the sea withdrew, the rising land mass was drained by streams that cut channels in the newly exposed shales. These channels were analogous to the distributary streams of the large deltas of today. Many of the channels became filled with sand, and these sinuous masses of sand are preserved as "channel sandstones," another characteristic of the Pennsylvanian sequences of rocks.

Desmoinesian Series. The cyclic alternation of marine and nonmarine environments was more intense and more frequent in the Desmoinesian than in other parts of the Pennsylvanian. Because the formations of the Desmoinesian are so numerous and so thin, and so repetitive in their composition and texture, it would be redundant to describe each one separately. Instead, the Desmoinesian column is shown graphically in Figures 6–23, 6–25, and 6–26. The Desmoinesian formations are divided into two groups, the Cherokee and Marmaton, and the total thickness is approximately 300 feet.

As seen in the drawings, the typical basal unit of a formation is a thin marine limestone that rests on the coal of the underlying formation. Over the limestone follows a succession of gray shale, then sandstone. Over the sandstone is usually a gray sticky clay called *underclay*. This clay commonly contains fossil plant remains. It is thought that this underclay represents the soil where swamp vegetation, the parent material of the coal, grew. The top of the coal is the base of the next formation.

The Desmoinesian contains most of Missouri's coal beds. Many are not of minable thickness, and some are not much more than black streaks in dark shales. However, they are important because they each represent a brief incursion of a swamp environment. Some range from a few inches to as much as a few feet in thickness. Mineral coal in northern Vernon County reaches a thickness of seven feet. The Tebo, which was once the most widely mined coal in the state, has a maximum thickness of three feet.

Missourian Series (Figs. 6–27 and 6–28). The Missourian differs from the Desmoinesian in that it has few coal beds and is composed almost entirely of alternating limestones and shales, both of which reach thicknesses much beyond the beds of the Desmoinesian. This fact suggests that the area became somewhat more stable during Missourian time.

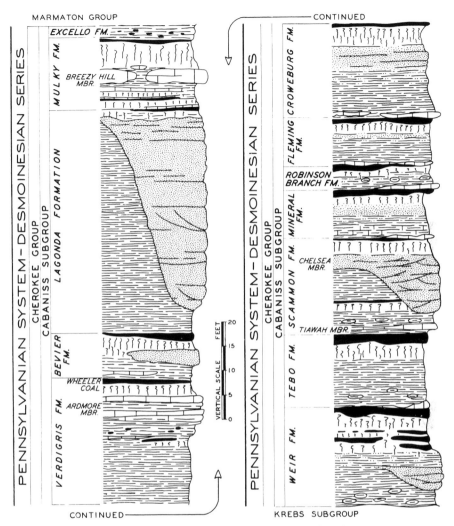

Fig. 6–25. Pennsylvanian System, Desmoinesian Series (Cherokee group, Cabaniss subgroup). DNR.

The fluctuations continued but were less frequent, and there were no more long-continued swamp conditions.

The limestones are mostly more massive and evenly bedded. They are abundantly fossiliferous with many bryozoa, brachiopods, and crinoids. There are some localized masses of algal remains. The fauna is distinctly different from that of the Desmoinesian; of particular interest is the appearance of the fusulinid genus *Triticites*, an indicator of Upper Pennsylvanian age (Fig. 6–29).

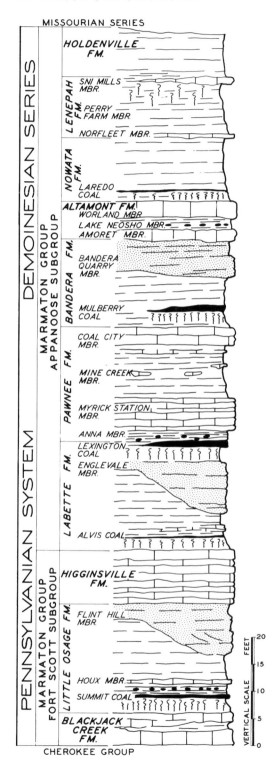

Fig. 6–26. Pennsylvanian System, Desmoinesian Series (Marmaton group). DNR.

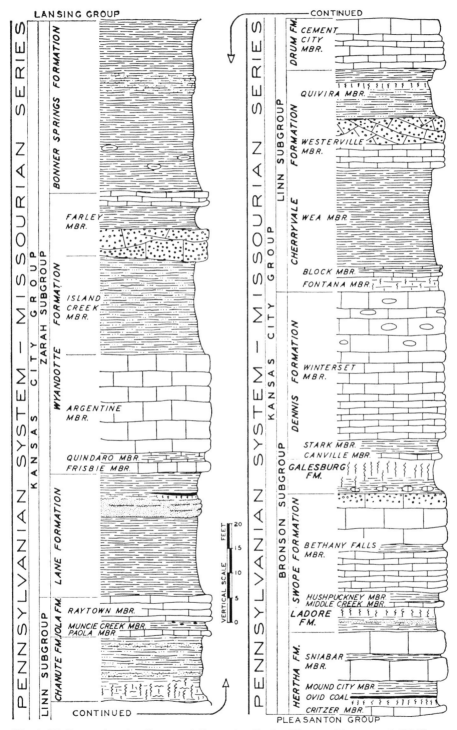

Fig. 6–27. Pennsylvanian System, Missourian Series (Kansas City group). DNR.

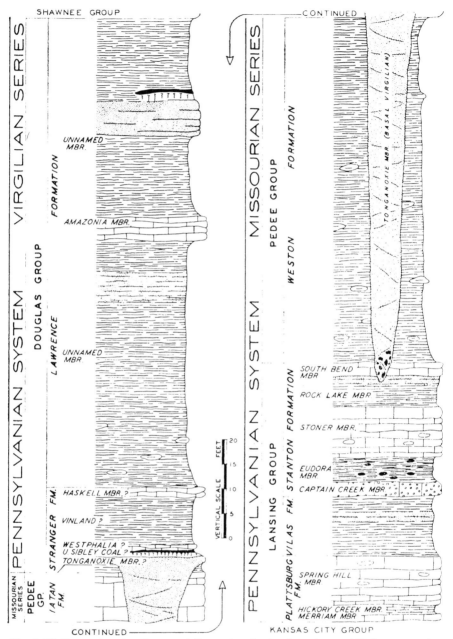

Fig. 6–28. Pennsylvanian System, Missourian Series (Lansing and Pedee groups) and Virgilian series (Douglas group). DNR.

Fig. 6-29. *Triticites,* a common microfossil used as an index for the Upper Pennsylvanian age of rocks. *Triticites* are also called "fossil wheat grains" because of their small size. Magnified fifteen times.

This series is divided into four groups, the Pleasanton, the Kansas City, the Lansing, and the Pedee, which are in turn divided into subgroups, formations, and members. Especially important among the limestones are the Bethany Falls Member of the Swope Formation, the Winterset Member of the Dennis Formation, and the Argentine Member of the Wyandotte Formation. These beds are quarried commercially for building stone, agstone, riprap (large, broken fragments used in such structures as dams, irrigations channels, and wing dikes), and road construction. In the vicinity of Kansas City they range from twenty to forty feet in thickness and form prominent ledges in the hillsides within the city (Fig. 6-30).

The shales of the Missourian Series are mostly gray and calcareous and carry many of the same fossils as the limestones (Figs. 6-31 and 6-32). The shales reach greater thicknesses than the limestones.

Sandstones are not common in the Missourian. One forms the base of the series in western Missouri, and it is called the Hepner. It is only about fifteen feet thick and is thinly bedded and micaceous. Another sandy zone occurs in the upper part of the Lane Formation. Two sandstones occur in the Pleasanton Group (not illustrated here). They are significant because they fill erosion channels that cut deeply into the two thick shales of the group. One, called the Warrensburg, represents a filling of what seems to be an extensive stream valley system that developed in late Missourian time, when the seas retreated from the area. This sandstone extends in a narrow elongate pattern southward from the Missouri River east of Lexington through Warrensburg to northern Henry County. It is exposed in road cuts along Highway 13 south from Higginsville. The sand is fine- to coarse-grained, micaceous, and strongly cross-bedded. It ranges from a few inches to as much as 150 feet in thickness. The Moberly Sandstone has similar characteristics and extends from southeastern Chariton County northeast across Monroe County.

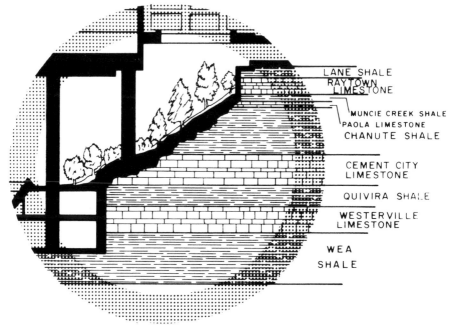

LANE SHALE
RAYTOWN LIMESTONE
MUNCIE CREEK SHALE
PAOLA LIMESTONE
CHANUTE SHALE
CEMENT CITY LIMESTONE
QUIVIRA SHALE
WESTERVILLE LIMESTONE
WEA SHALE

Fig. 6–30. The Crown Center Hotel in Kansas City features a waterfall and garden in its lobby. This unique interior garden was carved from alternating layers of Pennsylvanian limestone and shale in the hillside which were preserved as the hotel was built around them.

Virgilian Series (Figs. 6–33 and 6–34). The Virgilian Series outcrop area is smaller than those of the other series from the Pennsylvanian Period. It is limited to the area west of Caldwell, Daviess, and Harrison counties, north and east of the Missouri River. Like the Missourian Series the Virgilian is essentially free of coal and consists mostly of shale, limestone, and sandstone. There are four thin coals: the Sibley of the Stranger Formation, the Elmo, the Nyman, and the Nodaway.

Permian Period

The Permian, the most recent of the Paleozoic periods, is represented in Missouri by only about fifty feet of a massive, cross-bedded sandstone that occurs in a few localities in the northwestern corner of the state in Atchison County. The area of the outcrop is too small to be shown on the geologic map in this book. The sandstone occupies deep channels that are cut in the Upper Pennsylvanian beds, especially the Stotler Formation, and is called Indian Cave Sandstone for a locality in Nemaha County, Nebraska. It is early Permian in age.

Fig. 6-31. A portion of a slab of black Pennsylvanian shale from near Kansas City with two specimens of *Aesiocrinus*. Natural size.

Fig. 6–32. *Delocrinus missouriensis*, the crinoid that has been chosen as the Missouri state fossil. Natural size.

Interregional Correlations

The Geological Survey of America, in celebration of its centennial, is publishing a series of volumes that provides broad views and descriptions of the sedimentary cover of the North American craton (a *craton* is the central, stable region of a continent). An object of this effort is to recognize, on a time scale, the cratonwide contemporaneity of the thick sequences of sedimentary rocks and the periods of erosion and nondeposition (unconformities) that separate them. Geologists have determined not only that the processes of deposition are similar for the same time period over all of the craton but also that there are substantial unconformities that are common to the area.

L. L. Sloss (1963, 1982, 1988) has worked out time-stratigraphic sequences for the North American craton and applied names to major time units that include deposition and related contemporaneous erosion periods. These units are:

Sauk: from Late Precambrian to the border between Lower and Middle Ordovician

Tippecanoe: from top of Sauk to middle of Lower Devonian

Kaskaskia: from top of Tippecanoe to bottom of Lower Pennsylvanian

Absaroka: from Lower Pennsylvanian to the end of the Paleozoic

Bill Bunker et al. (1988) have detailed these events for the central Midcontinent region and related them to the Paleozoic rocks and unconformities in Missouri. Figures 6–35 and 6–36 show how the periods of deposition and nondeposition relate to each other throughout the craton, how they relate to the time periods as they are established

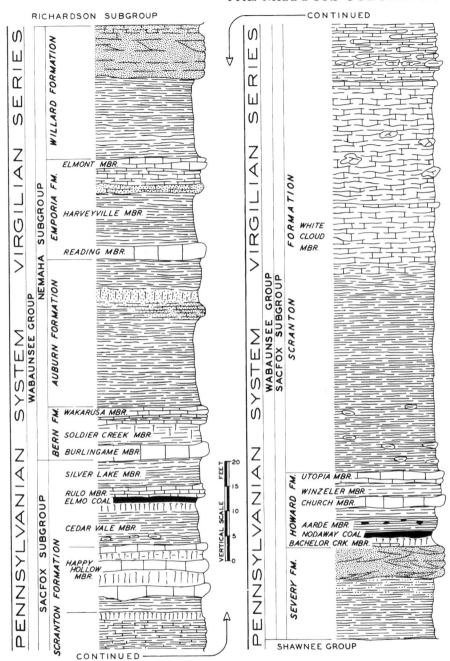

Fig. 6–33. Pennsylvanian System, Virgilian Series (Wabaunsee group, Sacfox and Nemaha subgroups). DNR.

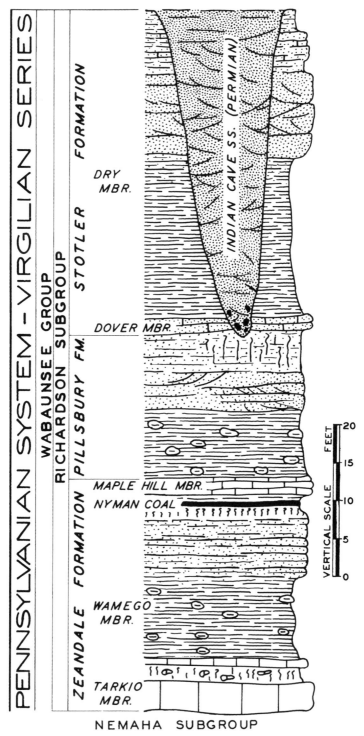

Fig. 6–34. Pennsylvanian System, Virgilian Series (Wabaunsee group, Richardson subgroup). The drawing also indicates a channel in the Stotler Formation that was filled in Permian time by the Indian Cave Sandstone. DNR.

QUATERNARY / TERTIARY	TEJAS
CRETACEOUS	ZUNI
JURASSIC	
TRIASSIC	
PERMIAN	
PENNSYLVANIAN	ABSAROKA
MISSISSIPPIAN	KASKASKIA
DEVONIAN	
SILURIAN	TIPPECANOE
ORDOVICIAN	
CAMBRIAN	SAUK

NONDEPOSITIONAL
DEPOSITIONAL
DEPOSITIONAL

Fig. 6-35. Idealized east-west section across the United States from the Cordilleran area to the Appalachians indicating times of erosion and nondeposition, versus times of deposition. After Sloss, 1988.

by geologists worldwide, and how they relate to the stratigraphy of the Midcontinent specifically.

Mesozoic Era

Only the Cretaceous Period of the Mesozoic Era is represented in Missouri (Fig. 6-37). In Ste. Genevieve and St. Francois counties there are exposures of igneous masses that are low in silica but high in iron and magnesium content. These are interpreted by some geologists as being diatremes that were emplaced during Cretaceous time. They seem to be similar to basic dikes known from drill holes in the Mississippi Embayment area and to Cretaceous intrusives in Arkansas. (A *diatreme* is a volcanic "pipe" formed by a gaseous explosion.)

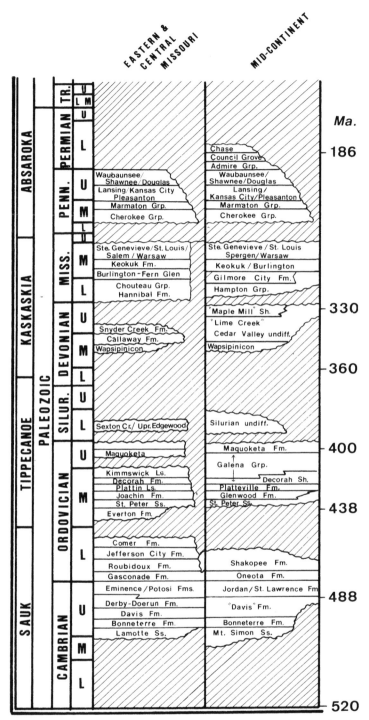

Fig. 6–36. Correlation of the Paleozoic rocks of Missouri to the standard Midcontinent time-stratigraphic scale. Ruled areas denote periods of erosion or nondeposition. Ma equals millions of years ago. Modified from Sloss, 1988, and Bunker et al., 1988.

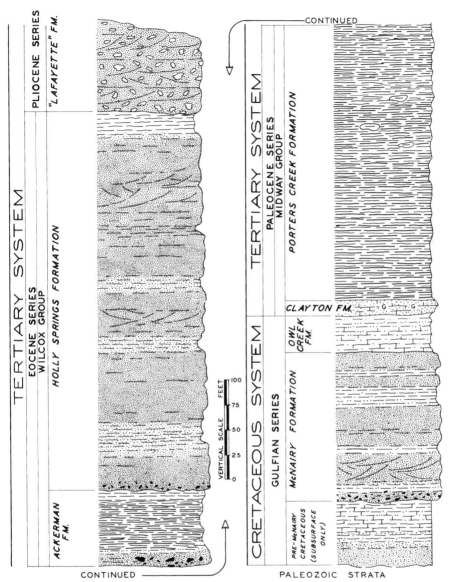

Fig. 6–37. Cretaceous and Tertiary systems. DNR.

A different kind of Cretaceous rock in southeastern Missouri was deposited by the northernmost extension of the Gulf of Mexico, that is, when Missouri was within the upper reach of the Mississippi Delta. The sediments deposited there are now exposed in Stoddard and Scott counties and represent only the Gulfian Series of the Upper Cretaceous, with a considerable unconformity at the top and at the bottom of the Gulfian. The Gulfian beds in southeastern Missouri are divided into three units, of which only two, the McNairy and the Owl Creek formations, are known at the surface.

Below the McNairy is a one-hundred-foot thickness of sand, chalk, clay, and limestone. Since these beds are known only from drilling and are neither well cemented nor hardened, their relationships in Missouri are uncertain. Some are marine and some nonmarine, and they can be considered comparable in age to the Coffee and Selma formations of Tennessee. Both beds dip to the southeast and thicken downdip (to *thicken downdip* means that the beds thicken as you go deeper).

Where seen in outcrop in Stoddard and Scott counties the McNairy Formation ranges from 100 to 250 feet and is composed of nonmarine sand and sandy clay. The basal part is mostly gravel, followed by thin layers of orange sand grading upward to a medium- to coarse-grained sandstone. The upper part of the formation is a succession of alternating beds of orange sandstone and clay. This formation is used as a commercial source of sand and is an important source of ground water.

The Owl Creek is almost completely a massive sandy micaceous clay. It is commonly glauconitic and carries marine fossils. Fresh exposures are bluish gray. The thickness ranges from eleven feet at the surface to one hundred feet downdip in the subsurface.

Cenozoic Era

As explained earlier, the term *Cenozoic* is used to designate the time since the Cretaceous. Even though it actually began some 65 million years ago, we think of it as including our most recent geologic events (ceno = recent; zoic = life). Most of the fossils found in rocks of this age can be closely matched with modern or very recent forms.

The Cenozoic rocks recognized in Missouri represent the Paleocene, Eocene, and Pliocene series of the Tertiary Period (see Fig. 6–37) and the Pleistocene Series of the Quaternary Period. And, of course, deposition of sedimentary rocks continues today. It makes geology more interesting to think of ourselves as being part of the whole of geologic history.

Paleocene Series. Two formations, the Clayton and the Porters Creek, compose the Midway Group. They are exposed along Crowley's Ridge in Scott and Stoddard counties but dip southward into the Mississippi Embayment, where they total more than 650 feet beneath Pemiscot County.

The Clayton Formation is only about ten feet thick in its outcrop area, where it is a fossiliferous calcareous sand and clay. It is characteristically green and bears glauconite. Some localities show limited amounts of limonite. The Clayton is conformable under the Porters Creek but lies unconformably upon the Owl Creek.

The Porters Creek Formation is a remarkably uniform, massive, dark gray to almost black clay. When exposed and dried, the gray becomes lighter. The formation varies in thickness, but in many places its outcrop exceeds 200 feet. In the thickest portions there are scattered masses of iron carbonate, and throughout its extent there are disseminations of white sand, mica, and gypsum. Much of the clay is bentonitic. It has some commercial potential because, when dry, it can absorb and hold a great deal of moisture. It is commonly processed to make "kitty litter." In well drillings, in the deeper sections, the lower fifty feet contain foraminifera and small clams. The Porters Creek is conformable on the Clayton but is overlain unconformably by the Ackerman Formation of the Eocene Series.

Eocene Series. Two formations, the Ackerman and the Holly Springs, represent the Eocene in Missouri. The Ackerman rests unconformably upon the Porters Creek Formation in Scott and Stoddard counties. In some localities where the Porters Creek is not present, the Ackerman overlaps onto the Cretaceous. In some places the material in the lower part of the Ackerman appears to have been derived by erosion from the Porters Creek. Most of the Ackerman is light gray to brown, silty nonmarine clay. It is in some places lignitic and bears glauconite in the basal layers. The upper part is plastic red clay. The formation is inconsistent in thickness and ranges from a few inches to as much as one hundred feet.

The Holly Springs Formation, above the Ackerman, varies greatly in composition from place to place. It is primarily a poorly consolidated sand that contains unevenly distributed clay and gravel. The sorting ranges from poor to excellent. Some portions are evenly bedded and some highly cross-bedded. The clay is sandy and silty and varies greatly in color from white through yellow, red, lavender, gray, green, brown, and black. The upper part in most places overlies an "iron-cemented" sandstone, and in one locality the upper part is chocolate brown. Some localities yield plant fossils. In the northwestern area of Crowley's Ridge the formation is only a few inches thick, but in the south end of the ridge it is more than 250 feet. At the base of the Holly Springs a bed of highly polished black gravel lies unconformably upon the Ackerman Formation.

Pliocene Series. The presence of Pliocene sediments in Missouri has been a subject of discourse among geologists for many years and still is not definitely settled. Throughout the Mississippi Embayment area of southeastern Missouri and in the adjacent Ozark region, the higher hills and intervalley divides are capped by layers of sand and gravel that range from a thin veneer to thicknesses of thirty feet near St. Louis and even sixty feet along the southeastern edge of Crowley's Ridge. These are often referred to as "high-level gravels." They lie upon an erosion surface that truncates beds ranging in age from Paleozoic to Eocene. Because these gravels occur uniformly over a wide area, they have been called the Lafayette Formation, named for similar gravels near Lafayette, Mississippi.

Most of the pebbles in the gravel are light brown, rounded, polished chert. Some are of quartz and some of quartzite. Although large cobbles and boulders are scattered throughout, most of the pebbles are one to three inches in diameter. In some localities they are admixed with medium- to coarse-grained, iron-stained sand.

Pleistocene Series. The name *Pleistocene* is applied to all but a small part of the most recent 2 million years. During this time, most of North America and northern Europe were covered at least four times by great continental sheets of ice. Thus, the Pleistocene is commonly referred to as the Ice Age. The last 10,000 years are referred to as the *Holocene.* However, we should not fail to realize that there were long times during which the ice disappeared from these areas. During these *interglacial intervals,* soils formed, plants grew, and animals roamed the countryside. Too often this aspect of the Pleistocene is overlooked. It is possible that what we call the Holocene may even now be one of these warmer intervals.

As they moved southward from the frozen North, glaciers scoured the underlying landscape and gathered within the ice much rock material. As the ice melted, this material, consisting mostly of clay, sand, and gravel, settled in the form of a widespread sheet collectively called *glacial drift,* or *till.* Upon this layer of material, soils formed during the interglacial intervals. These soils are used by geologists to subdivide the glacial times. In the United States geologists have for many years used the following designations for the four major ice sheets and the interglacial intervals.

Recent	since the melting of the last glacial ice
Wisconsinan	glacial
Sangamonian	interglacial
Illinoian	glacial
Yarmouthian	interglacial
Kansan	glacial
Aftonian	interglacial
Nebraskan	glacial

The interglacial soils vary from place to place in color, texture, and thickness, and in some places it is almost impossible to tell the older from the younger. The maximum thickness of the deposits at any one place is about 400 feet. Over most of northern Missouri they conceal the underlying preglacial surface and the older bedrock features. These deposits are important sources of sand and gravel. Where the glacial drift is highly permeable, it contains significant amounts of ground water. Also of great importance is the fact that the drift provided the parent material for the rich soils of northern Missouri.[9]

In addition to the widespread layers of clay, sand, and gravel, the glaciers left behind large boulders of igneous and metamorphic rock not native to Missouri that were carried by the ice for distances of over 300 miles. They are called *erratics*.[10] The largest and best preserved is known as the Bairdstown Church Erratic and lies about eight miles northwest of Milan (Fig. 6–38). The exposed part of this pink granite boulder is a rounded prism about twenty feet long, twenty feet wide, and eight feet high. It is estimated to have weighed 384 tons.

In Adair County two large boulders were found about two miles southeast of Novinger. One is a red granite mass about seven feet long and six feet in diameter. It now marks the site of Fort Clark and "The Cabins," the first settlement in Adair County. The other is a dark gray granodiorite about three feet in maximum diameter. It is now on the Northeast Missouri State University campus in Kirksville.

Four giants are known in Carroll County. One is red quartzite, measuring twelve, by nine, by three and a half feet. The second is red granite and about five by six feet; the third is quartzite, about five feet square; and the fourth is of andesite and about six feet square.

During the excavation for the Geology Building on the Columbia Campus of the University of Missouri in 1964, a huge boulder of granodiorite was encountered about five feet underground. It was drilled

Fig. 6-38. The Bairdstown Church Erratic, near Milan, is the largest and best preserved erratic known in northern Missouri. This red granite boulder was probably moved at least 300 miles by the ice. DNR-Vineyard.

and split. About half of it, in four pieces, is now on the edge of the parking lot behind the building. The pieces range in size from about two feet to over five feet in largest dimension. The other half of the boulder is still buried.

Another large red granite boulder marks the grave site of G. C. Swallow in the Columbia Cemetery (Fig. 6-39). Swallow was the first state geologist of Missouri and the first dean of agriculture at the University of Missouri. The boulder was found on a farm about three miles west of Columbia and in 1928 was moved by a group of Columbians who wanted it to serve as a memorial to Swallow.

It is not unusual, when traveling through northern Missouri, to see glacial erratics displayed in yards or gardens. The largest collection known to the authors, about seventeen stones, is displayed around the parking area for Nodaway Island Access on the Missouri River near the village of Nodaway in Andrew County. The river access is owned by the Missouri Department of Conservation, and the erratics were collected from farms in the surrounding countryside. The sizes of erratic boulders and the distances they were moved by glaciers give an impressive indication of the size and power of the ice sheets.

The glacial deposits have also preserved for us a fascinating record of the vertebrate life of the time. Skeletal remains of a variety of Ice Age animals have been found across the state in gravel pits, river valley al-

Fig. 6–39. This glacial erratic marks the grave of G. C. Swallow in the Columbia Cemetery. It was first found in a field about three miles west of Columbia and was moved to its present site in 1927.

luvium, sinkholes, and caves (Fig. 6–40).[11] Among the most common animal groups represented are the mammoths, mastodons, horses, sloths, peccaries, musk-oxen, and bison. More rare, but present, are tapirs, deer, camels, and bears. Other species include rabbits, armadillos, foxes, squirrels, rats, mice, turtles, and snakes. In addition, trackways and a tooth of a saber-toothed tiger were found in a large cave near Perryville, and tracks of an Ice Age lion were found in a cave in south-central Missouri. In just one sinkhole in eastern Missouri, ninety different animal species were identified.

Probably the most famous Ice Age fossil found in Missouri is the one that Maurice G. Mehl (1966) christened the Grundel Mastodon. This fragmented skeleton was found in a gravel pit on the farm of Russell Grundel in Holt County in 1963. It was buried in the Farmdale Loess of Wisconsinan age. It was dated at about 25,000 years before the present. Although mastodon remains have been found in 33 of Missouri's 114 counties (Fig. 6–41), most of the finds are of single fragments of teeth or bones. The Grundel find is unusual in that many fragments—teeth, four ribs, a lower jaw, six vertebrae, and a large tusk—all came from the same animal. Another exceptional feature of this find is that a nearby "fossil hearth" was judged by archaeologists, on the basis of radiocarbon dating of the charcoal, to have been built approximately 25,000 years ago. On this evidence Mehl postulated that there was human activity contemporaneous with the death of the mastodon. The tusk is on display at the

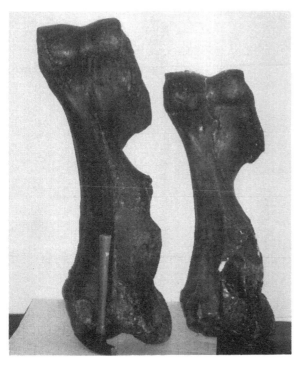

Fig. 6–40. Upper leg bones from Pleistocene mammoths found in Boone County.

Division of Geology and Land Survey in Rolla, along with vertebrae, ribs, and skull parts from a mammoth found near Troy, Missouri.

One of the first "finds" of bones of mastodons and other Ice Age animals was in Jefferson County near Imperial. Here Albert Koch, a German naturalist, excavated bones and teeth of many large extinct animals in 1839. Koch assembled one large skeleton, which he named *Missourium Theristocaulodon* and nicknamed the "Missouri Leviathan" (Fig. 6–42). He also reportedly sold one large skeleton to the British Museum of Natural History in London. This assemblage of bones was regarded by many scientists as an exaggeration. Koch reported that it was thirty feet long and fifteen feet high, but the vertebral column was extended by the insertion of bones belonging to many individuals. The tusks were described as curving out horizontally. Koch went so far as to say that this animal was the one described in the Bible in the forty-first chapter of Job. He also suggested that the animal's death was caused by "a certain comet that came in contact with our globe."[12]

The Jefferson County locality was more or less ignored until 1897, when C. W. Beehler renewed digging and attracted widespread interest. He is reported to have uncovered more than sixty mastodons in addition to many other vertebrates. Unfortunately, these specimens were

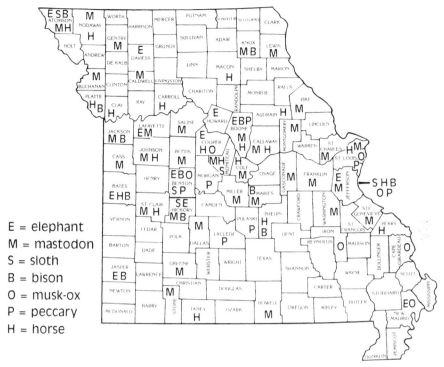

E = elephant
M = mastodon
S = sloth
B = bison
O = musk-ox
P = peccary
H = horse

Fig. 6–41. Fossil finds in Missouri by county.

not cared for and are lost to science. In 1970 the citizens of Jefferson County decided to protect the area and gathered support from the public and the state. As a result, the Missouri Department of Natural Resources purchased 418 acres to create Mastodon State Park. The visitors' center there offers exhibits of bones, teeth, tusks, and other remains of Ice Age creatures, including an anatomically complete and correct replica of a mastodon skeleton that was assembled by paleontologists of the Illinois State Museum.

Although no ice sheet invaded Missouri after the Illinoian, other geologic processes were at work that left evidence in the rocks and other geologic features of the state. Over much of the state, and particularly on high ground along the Missouri and Mississippi rivers and some of their tributaries, are extensive layers of medium- to coarse-grained, light to dark brown silt called *loess*. Loess contains some clay and in some places is calcareous, but it varies greatly from place to place. Many localities yield small fossil snail shells. The thickness of the loess varies, with the greater thicknesses being closer to the river and on the eastern

Fig. 6-42. Albert Koch's *Missourium Theristocaulodon,* or the "Missouri Leviathan." From Merrill, 1924.

side. The thickest known loess in Missouri is 122 feet, but thicknesses of from 10 to 15 feet are more common.

The origin of this silty material has been debated, but many features suggest that silt from the river floodplains is blown to higher levels by strong winds during periods when the plains are exposed and dried. This process can sometimes be observed on windy days along the Missouri River. In contrast to the way most unconsolidated sediments slump to form gentle slopes when exposed by excavations such as road cuts, the loess forms and holds nearly vertical cliff faces (Fig. 6-43).

*   *   *

In Chapter 2 we talked about learning the "language of the rocks." After our examination of the Missouri column, it should be clear how the character of the sedimentary rocks can tell us the geologic history of an area—how it has looked and how it has changed over time. We can determine whether an area was covered by the sea, was the site of a coal swamp, or perhaps was barren and subject to weathering and erosion. If it was covered by water, we can tell if the water was warm or cool, muddy or clear. We can tell when the earth's crust warped and when the sea level fluctuated.

As we read the rocks, we become aware that over millions of years, the earth undergoes great changes. Mountains are raised and worn away. Oceans move from place to place. So do large continental seg-

Fig. 6-43. A road cut through thick layers of loess near Rockport, Missouri, illustrates how this material maintains a vertical cliff face. DNR-Vineyard.

ments of the earth. Great ice sheets form and move across continental regions, then melt, to be followed thousands of years later by other ice sheets. The story of all these events is recorded in the rocks. The column of rocks and sediments in any location is a time line of that location's geologic history. If one can read the appropriate signs in that column, one can read that history.

7

MINERAL AND GEOLOGIC RESOURCES

However one looks at the history and economy of Missouri, it becomes obvious that the geology of the state is, and has always been, a major factor in the state's history and development. Its rivers and their tributaries were migration routes for early man and for the pioneer settlers. The junction of the Missouri and Mississippi rivers provided an ideal setting for the establishment of St. Louis. That city then became a point of departure for many early explorations using the Missouri River to reach areas in the West and Northwest. The confluence of the Missouri and the Kansas rivers, in a similar way, gave rise to Kansas City.

The cities of Joplin, Webb City, and Carthage developed around the lead and zinc mines of southwestern Missouri. In the eastern part of the state, lead and iron ores have been mined from the St. Francois Mountains since their earliest white settlement in the eighteenth century. Discoveries of iron ores in the margins of the St. Francois Mountains brought about the development of an iron and steel industry that boosted the economy of early St. Louis. Mining of iron ore continues today at the Pea Ridge iron mine near Sullivan. The massive granite bodies in the same mountains provided paving blocks for the levees and streets of early St. Louis. Abundant supplies of clay and coal led to the development of a brick industry in central and east- central areas of the state. These resources provided the raw materials for the construction of commercial buildings and of stately brick mansions for the elite society of early St. Louis; many of these mansions still exist.

The state has been a leader in mineral production for more than 250 years. In 1990 Missouri ranked tenth in the nation in nonfuel commodity values. In 1990, Missouri ranked first nationally in production of fireclay, lime, and lead; second in the production of iron oxide pigments; third in barite and iron ore; fourth in zinc; fifth in portland cement and fuller's earth; and sixth in copper. These are impressive rankings in view of the fact that Missouri has a population of 5 million, only 1.6 percent of the nation's population; many of these commodities and products are shipped and sold elsewhere, bringing dollars into Missouri's economy.

Within the state, lead ranked first in value at $390 million, crushed stone second at $191 million, and portland cement third at $170 million. Lime, copper, fuller's earth, and iron oxide pigments totaled $202 million. Zinc production was valued at $85 million, with silver at nearly $7 million. The total value of mineral production in Missouri during 1990 was reported by the U.S. Bureau of Mines to be $1,110,302,000. When the value of products manufactured directly from raw materials produced in Missouri's mines is added, the total jumps to more than $2 billion. Taking into account salaries, taxes, supplies purchased, freight, and royalties, the industry generates almost $4 billion into Missouri's economy.

The balance of this chapter will consider each of the state's important resources, providing further information on the geology, production, and value of each one. Barite, cement, chert, clay and shale, coal, copper, gemstones, iron ore, lead, lime, oil and gas, sand and gravel, silica sand, silver, stone, tripoli, tungsten, water, and zinc will all be examined in alphabetical order. For most of these resources, Plate 5 indicates the areas of the state in which they can be found.[1]

Barite

Barite, commonly called *tiff,* is not well known to most people, even though it has been mined in Missouri since 1850, and at times the state has been the leading producer in the United States. In the early days of pick-and-shovel lead mining, barite was discarded as waste matter, which is apparently the reason it was called tiff. As industrial uses for it were developed in the Old Mines area of Washington County, where it was abundant, it became important, and beginning about 1850 it was gathered by hand and sold for only two or three dollars a ton.

Barite is barium sulfate to the chemist and is, in an ordinary sense, chemically inert. It is a gray to white, relatively soft mineral that is exceptionally "heavy," about four and a half times as heavy as an equal volume of water. Because of this high specific gravity, its most common industrial use is as a filler with the water and clay minerals that make up oil-well drilling mud. This drilling mud is pumped through the drill system to cool the bit and keep the hole from collapsing. Because of its "weight," barite helps the mud carry the lighter cuttings to the top and helps control high pressures in deep wells. Other uses for barite are as a filler in paint, rubber, glass, printing ink, paper, and textiles. The coating that creates the slick appearance of magazine paper is an example. Also, because of its density, it is used as an additive in concrete aggregate in the foundations and walls of nuclear reactors, where it serves as a barrier to radioactive leakage.

Most Missouri barite has been produced in Washington, Franklin, Jefferson, Cole, and Moniteau counties. Small amounts have come from Cooper, Morgan, Miller, Camden, Benton, Hickory, and St. Clair counties. Maximum production was reached in 1956 with 381,642 short tons valued at $4,461,955 (for the distinctions among short, long, and metric tons, see Weight Measurements in the Glossary). Production in 1989 was 18,000 short tons valued at $969,000. Production in Missouri has been declining because of a slump in exploratory drilling for oil in the United States and increasing barite production in other states.

In Washington County and the surrounding area, most of the barite is in red clay residual deposits derived from the weathering and removal of rocks from the Potosi and Eminence formations of the Cambrian Period, where the mineral was deposited in veins by mineral-rich solutions. Small quantities also occur as veins in the Ordovician Gasconade, Roubidoux, Jefferson City, and Cotter formations, where it was deposited by mineral-rich solutions moving through the host rocks.

Cement

The material commonly referred to as cement has long been, and still is, the major mineral commodity produced in the United States, and it is likely to continue this high ranking. In Missouri cement has, for more than thirty-five years, ranked in first or second place in value and production; nationally, the state has ranked sixth in production for many years, climbing to fifth in 1990. Cement is properly referred to as *portland cement,* so named in 1842 by the Englishman, Joseph Aspdin, who obtained a patent for its manufacture, because his product resembled stone quarried on the Isle of Portland. The term *cement* is commonly misused to designate concrete, which is a mixture of cement with an aggregate material.

Cement is made by blending limestone, clay or shale, and sometimes sand, in the proper proportions; burning the mixture to 2,500–3,000 degrees F to form what is called a clinker; adding gypsum; and then pulverizing all to a fine powder. When the fine powder is mixed with water and allowed to "set," a rock-hard solid results. If an aggregate, such as crushed limestone or gravel, is mixed in at the same time as the water, the product is concrete, the most widely used building material in the United States.

The first cement production in Missouri began in Ralls County in 1901, and the industry has grown greatly in counties along the Mississippi River south to Cape Girardeau County. In addition, one plant operates in Jackson County. The major factor determining the location of the plants is the availability of the appropriate limestone and shale. Another factor is the proximity of the Mississippi and Missouri rivers, for economical water transportation. The raw materials needed are abundant in the state, and Missouri can support the present industry and meet future expansion. It has been estimated that annual production here will probably remain about 30 million barrels well into the 1990s. Production of portland and masonry cement in 1990 was 4.6 million short tons valued at $170,200,000.

Chert/Flint

"A rose by any other name is still a rose." So it is with chert. Whether called jasper, agate, flint, or mozarkite, it's still chert. Chert is not usually listed as one of Missouri's valuable mineral resources, and statistics of its production and dollar value are not collected and published. Nevertheless, it is one of the state's most abundant and widely known rock products.

The early human inhabitants used flint for arrowheads, spear points, skin scrapers, awls, stone axes, grain-grinding pestles, and many other tools and implements (Fig. 7–1). Today many tons of chert are used for road surfacing in the form of loose gravel or in the asphalt used for "blacktop." It is popular for this use because it wears well and is abundant and easily obtained as stream gravel. Because chert is relatively insoluble, it remains behind as the chert-bearing limestones and dolomites dissolve, accumulating in streams or as lag on rock-covered hillsides.

Chert occurs in almost every carbonate formation in Missouri; it occurs as nodules, beads, lenses, stringers, or in some formations as distinct beds. Concretions take various forms and are sometimes mistaken for fossil bones, fossil melons, or petrified eggs. Chert is dominantly composed of interlocking crystals of quartz (SiO_2) less than thirty microns in diameter. Most chert is white or light gray, but frequently it is colored in various shades of red, tan, brown, and yellow; some may be grayish blue or black. Mozarkite, declared the "official rock and litho-

Fig. 7–1. Indian artifacts made from flint from the Burlington Formation.

logic emblem of Missouri" by the 74th General Assembly, has various shades of gray, red, or pink. Agate may be banded in various colors interlayered with white.

Differing theories have been proposed about the origin of chert. Some geologists believe that the deposition of silica occurred at the same time as the surrounding material accumulated. Others believe the silica was later transported into the carbonate rocks by silica-rich ground water and deposited in pore spaces or in solution cavities. Accordingly the age of any vein of chert is indefinite; it may be the same as the enclosing formation, or younger.

One undesirable characteristic of chert keeps it from being useful as a concrete aggregate. When fresh chert is exposed to water and to the chemicals present in cement, it becomes chemically hydrated and swells to become a hydrous gel. If subjected to freezing, this process is exacerbated; over time, this causes concrete to disintegrate.

Clay and Shale

For simplicity and convenience, clay and shale are considered here as one commodity, although they are not quite the same. Combined, they have long been rated as the eighth most valuable mineral product in Missouri. Bits and pieces of broken pots and vases found in old Indian villages and campsites tell us that clay was a valuable resource to the earliest human inhabitants of the state. The old brick chimneys and brick homes built by the early French settlers along the Mississippi are also examples of the early use of Missouri clay.

Shale is a sedimentary rock composed of one or more clay minerals mixed with considerable quantities (as much as 50 percent) of silt-size

(.004–.05 mm) particles of other minerals (geologists consider particles larger than .004 mm to be silt, while soil scientists use .002 mm and engineers use 0.005 mm). Shale is typically laminated or stratified.

Clay is defined by geologists as consisting of particles measuring less than $1/256$ mm (.004 mm). Clay shows no tendency to be stratified and is typically structureless. Its minerals form by the physical and chemical alteration of previously existing rocks, and the nature of the clay is of course affected by the mineral composition of the original rock. The clay minerals tend to form very small crystalline particles, and in the weathering and transportation process they may become mixed with other materials. Commonly they become associated with silt-sized fragments and are transported into quiet water before deposition. When this happens, the result is most often laminated shale beds. However, if the clay particles become separated and deposited without the mixture, the result will be a mass of nonstructured, nonlaminated pure clay. Even then some alteration may still take place under the influence of ground water or by compaction under pressure. Many of Missouri's high-grade clays have been formed in this way by being washed into old sinkholes where they have been altered to become refractory-quality clay.

This refractory clay, or fireclay (see Fig. 1–6), is the most valuable clay in Missouri and is distinguished by having a very high fusion temperature. The best fireclays fuse only at temperatures above 2,775 degrees F. This characteristic is controlled by the alumina (aluminum oxide) content, which in high-grade refractory clays exceeds 60 percent. Fireclay is used for production of material that will resist high temperatures, such as linings for furnaces in foundries and in steel manufacturing.

W. D. Keller (1979) describes the history of a valuable refractory clay called *diaspore*, which had an alumina content above 65 percent. It was discovered about 1917 and was produced in six east-central counties—Osage, Gasconade, Franklin, Maries, Phelps, and Crawford—until about 1950 when all workable deposits were depleted. Keller's story relates the "discovery, waxing, peaking, waning and depletion of a non-renewable resource, all occurring within one generation."

In 1990, as for many years, Missouri led the nation in production of refractory clay. Today, this important material is present in three areas in the state: the northern district (Audrain, Boone, Callaway, Monroe, Montgomery, Randolph, and Warren counties); the southern district (Crawford, Franklin, Gasconade, Maries, and Phelps counties); and the St. Louis area. All of the clay in these areas is Pennsylvanian in age. The northern district is one of the largest refractory clay districts in the United States, and there are important factories in Mexico, Missouri, and Fulton. The rocket-launching pads at the Kennedy Space Center at Cape Canaveral were paved with blocks made at the A. P. Green factory in Mexico.

Other varieties of clay and shale also have their particular uses and economic value, and in 1990 the combined production of common and refractory clay and shale was 1,426,925 short tons valued at $13,594,000. As mentioned in the section on cement, the abundant shales of the Paleozoic, particularly those from Pennsylvanian, Devonian, and Middle Ordovician times, have been essential to the development of Missouri's cement industry. Cement manufacture uses about 45 percent of the state's total clay and shale production.

A nonswelling bentonitic clay found in Scott and Stoddard counties is used for absorbent and carrier uses. One of its biggest uses now is in the manufacture of kitty litter. This clay occurs in the Paleocene Porters Creek Formation.

The nonrefractory clays and shales have varied uses as building bricks, face bricks, drain tiles, roofing tiles, terra-cotta, flowerpots, other pottery, and stoneware. Before the great increase in use of concrete and asphalt, many streets and highways were paved with bricks. Some clays have the ability to bloat when heated to high temperatures, and these are used to make a lightweight aggregate used in concrete building blocks. The "haydite block" is one of these products.

Kaolinite, the clay most commonly used for making white china or porcelain items, is also used as a filler and a strengthener in making paper. It is especially important in providing a smooth surface for the glossy paper used in printing many magazines and other items.[2]

Coal

There are more than forty separate coal beds in Missouri, and coal occurs in sixty-three counties and beneath more than 23,000 square miles in the northwest, west-central, and north-central parts of the state. However, not all the coal is minable or commercially valuable. Only about half of the beds are of minable thickness (over fourteen inches), and since 1925 essentially all of the mining has been done by surface methods (Figs. 7–2 and 7–3). In 1967 Walter Searight reported that less than 1 percent was then being mined underground. The most important coals are in the Desmoinesian Series. They are ranked as "high volatile bituminous."

Most of the coal production has come from five major fields (Fig.

Fig. 7–2. A coal strip mine in the Weir-Pittsburg coal in Barton County.

Fig. 7–3. Croweberg coal as seen in the highwall of a strip mine in Vernon County.

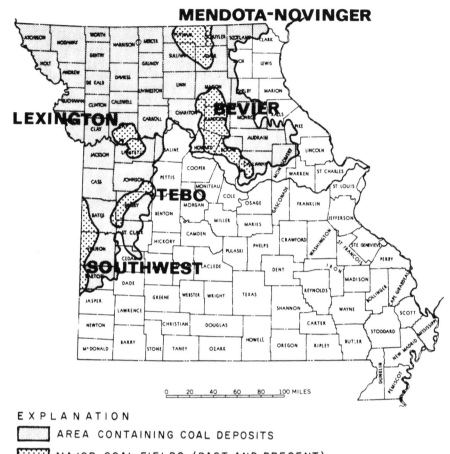

Fig. 7-4. Areas of coal deposits in Missouri and major coal fields (past and present). From Robertson, 1973.

7-4): Southwest Field, encompassing portions of Bates, Vernon, and Barton counties; Tebo Field, in Henry, Johnson, and northeastern Bates counties; Lexington Field, in central and north-central Lafayette County and southern Ray County; Bevier Field, the largest field, which extends from Callaway County northwestward across Boone, Howard, Chariton, and Randolph into Macon County; and Mendota-Novinger Field, mostly in Putnam and Adair counties.

Only about one-third of the coal used in Missouri is mined here. Because the St. Louis area is so near the Illinois coalfields, much of the state's need is supplied from there. The growing population of western Missouri consumes most of the production from the western coalfields and requires additional supplies. Currently most of the coal being used in western Missouri is low-sulfur coal from Wyoming.

Missouri is ranked ninth among the twenty-seven states with reserves of bituminous coals, according to the Division of Geology and Land Survey, and fourteenth among thirty-two states in total coal resources. Records on annual production since 1840 are available,[3] and the total to 1967 was 309,256,840 short tons. The peak year was 1917 with a total of 5,670,549 tons. In 1987, production was 4,085,000 short tons valued at $112,338,000. Charles E. Robertson (1973) estimated Missouri's total coal resources at 49 billion tons, which is ample to support an extended coal mining industry. He also reported that no significant areas of low-sulfur coal were present in Missouri, though some small reserves of low-sulfur coal have been found recently.

Cobalt and Nickel

Missouri production of cobalt and nickel has been negligible, but older literature mentions early attempts to produce these important materials. Old Mine La Motte produced nickel at various times in the nineteenth and early twentieth centuries, but only in small amounts that did not justify further development. Cobalt and nickel are present in the lead ores of the southeastern region, but not in sufficient quantities to make recovery profitable until market prices increase considerably, or until industry develops an economical method of extraction.

Copper

Although the United States has been the world's leading copper producer, Missouri's contribution is very small, only about one-tenth of 1 percent of the total. It often comes as a surprise to find that Missouri produces any copper at all. However, Missouri copper has been produced as a by-product in lead and zinc mining, from ores in which copper entered the pore spaces of the host rock in mineralizing solutions along with lead, nickel, and cobalt. Some copper minerals were produced in the Shannon County area prior to the 1920s, and lead mining in southeastern Missouri has led to increased production of copper as a by-product.

Although the annual value of copper production in Missouri is small relative to other metals, it ranged above $1 million in the 1950s and again in the sixties, reaching $2.5 million in 1967. The eighties saw greater increases, and in 1983 production reached a value of slightly more than $13 million, but it dropped to $8.5 million in 1984. In 1990 Missouri ranked sixth among the states in the production of copper.

Gemstones

Gemstone production in Missouri has never reached significant monetary value. However, because of the great variety and wealth of minerals and rocks, the state has long had active gem and mineral clubs, and various gem and mineral specimens have been collected since early man came into the area.

Most of the gemstones found here are not of the "precious" category, but sometimes their worth is related to personal preference or sentimental values. Quartz-family materials predominate—agate, jasper, and smoky quartz especially. Quartz is an abundant natural mineral; its components, oxygen and silicon, are the two most abundant materials in the earth's crust. Quartz-lined geodes and quartz druse specimens are plentiful in some localities and can be found on sale in roadside stands in the Ozark area. Beautiful crystals of other ore minerals may also be found—among them barite, calcite, fluorite, galena, hematite, pyrite, marcasite, and sphalerite—but they are not common.

Stones are also used for ornamental purposes. A pink chert that takes a fine polish occurs mostly in the Cotter Dolomite. This is the stone called mozarkite, which is used for belt buckles, necklaces, and bolo-tie clasps.

Iron Ore

Iron is the second most abundant metallic element in the earth's crust, after aluminum, and it unites readily with other elements to form many minerals. It can be extracted economically from most of them. Although Missouri has never played a major role in the national industry, the production of iron ore has been important to Missouri's economy for many years.

The most important source of iron ore in the state has been, and still is, the Precambrian core area of the Ozark Uplift in Iron County, which has been producing ore since 1815. The igneous rocks contain magnetite and other iron-rich minerals that were crystallized as the original magma cooled. When these minerals weather, the iron becomes oxidized and forms the more stable minerals hematite and limonite. Early in the state's history, near-surface mines at Iron Mountain and Pilot Knob were the largest producers of hematite. As the near-surface ores were worked out, surveys made by towing magnetometers behind airplanes guided mining companies to buried ore bodies rich in magnetite. Among these discoveries are ores now at Pea Ridge and at Pilot Knob. Others not yet developed occur at Bourbon, Boss, Kratz Spring, and Camel's Hump.

Two other types of ore that have been important in the past will be less so in the future. One of these is referred to as "filled sink" hematite, because it occurs in old sinkholes in Cambrian and Ordovician rocks (mostly in the Gasconade Formation) on the northwest flank of the St. Francois Mountains. There, the iron-rich sediments derived from the weathering of the igneous masses were deposited. This ore is mostly red to purple hematite with smaller amounts of blue specular hematite and small amounts of limonite. About 75 percent of the deposits are in Crawford, Phelps, and Dent counties. There has been very little production since the end of the nineteenth century.

The other ore of lessening importance is referred to as "brown iron ore," and it occurs throughout the Ozark region. However, most of the deposits are small and widely scattered, and the grade varies greatly from one location to another, and even within a single location. These ores are mostly limonite and occur in irregular masses and pockets; some have an undesirable content of sulfur. They are not amenable to processes that would increase their value, and there is little possibility that demand for these ores will ever increase.

Only the Precambrian ores have exceptional future potential. They are reported to be of sufficient extent and size for long-range production and to be amenable to conversion to the high-grade pellets used in blast-furnace operation. Especially important to the future is the Pea Ridge Iron Mine near Sullivan, the only underground iron mine in the nation. The Bourbon deposit is likely to be the next major development.

The first iron furnace operated west of the Mississippi River was built near Ironton in 1815. Small furnaces were also put in operation near Bourbon and Caledonia, and by 1823 the three chief types of iron ore—magnetite, limonite, and hematite—were being exploited. Other more important works were established at Maramec Spring in 1829, at Iron Mountain in 1846, and at Pilot Knob in 1848. In 1887 iron production reached a peak of 123,700 tons. However, the industry declined and production of so-called charcoal iron, or iron produced in a charcoal-fired furnace, ceased in 1925.

The decline in the iron-making industry in Missouri did not stop the mining and production of the ore, which was still used by iron and steel works elsewhere. The Granite City Steel Plant in Illinois across the river from St. Louis is the prime user of the pelletized ore produced by the Pea Ridge operations. Since 1963, Missouri iron ore production has averaged more than a million long tons annually. In 1984 ten states were producing iron ore, and Missouri ranked third with about 1.9 million long tons, all from Pea Ridge. Production in 1989 was 1,118,000 metric tons, but it declined slightly in 1990 to 947,000 tons.

Lead

If it had not been for the lead and zinc ores of Missouri, the history of the state and of the Midwest would have been very different. The early French explorers discovered lead ore in what is now Madison County in 1701, and the famous Mine La Motte opened in 1720. Other deposits were found near Potosi in the late 1700s and in what is now Washington, St. Francois, Franklin, and Jefferson counties in the early 1800s. Lead was discovered within the city limits of Joplin in 1848, and mining started in Granby in 1850. Missouri has been a leading producer ever since.

T. H. Kiilsgard (1967) has said, "Lead ore has been mined from hundreds of deposits in southeast Missouri," although many of these deposits are small and not of great importance. The lead-mining areas of the state can be divided into three districts: the Southeast District, the Central District, and the Southwest District.

The Southeast District

The "Old Lead Belt" includes the area around Flat River in St. Francois County and the Fredericktown area in northern Madison County. The latter includes the famous Mine La Motte. This "belt" was for many years the biggest and the most important producer of lead in the state. Most of the ores in this area were less than 400 feet below the surface. In addition to lead they yielded significant by-products of zinc, copper, nickel, and cobalt.

Foreseeing the end of the minable ore deposits in the Old Lead Belt, mining companies began exploring outward and downward, trying to find new deposits to replace those about to be mined out. The Viburnum Trend, sometimes called the "New Lead Belt," extends from southern Crawford and Washington counties across Iron County into Reynolds County. The initial discoveries here were made in the late 1950s by the St. Joseph Lead Company, but other companies have expanded the district. In 1984 the largest single producing unit in the United States was the Buick Mine. The second was the Magmont Mine. Both of those are in Iron County. The third largest was the Fletcher Mine in Reynolds County. In the Viburnum Trend, the ores occur at different levels between 1,000 and 1,450 feet below the surface.

Since the first ore was brought to the surface in the Viburnum Trend in 1960, the deposits have yielded more than 9 million tons of lead and large amounts of other metals, including zinc, copper, and silver. Thanks to advances in mining technology, the Viburnum Trend mines produced more metals in thirty years than were brought to the surface in the Old Lead Belt in the 107 years of mining from 1865 to 1972.

The Viburnum Trend is a excellent example of ore deposits that are explored for, discovered, mined, and closed down after mining has been completed. The first Viburnum Trend mine, Indian Creek, is already closed; the remaining mines are all scheduled to shut down within the next ten to fifteen years, according to the closure plans filed to comply with the metallic minerals waste law. As yet, no successor to the Viburnum Trend has been announced, although north of the Trend, in Washington County, there have been small workings around Palmer.

Together, the deposits of the southeastern district form a roughly circular pattern around the Precambrian formations of the St. Francois Mountains. Most of the ore is in the Cambrian Bonneterre Formation in areas where porous zones have been permeated by the lead-bearing solutions. The precise mechanism for introduction and deposition has been under study for many years. The area has become a classic, and geologists refer to the lead here as Mississippi Valley–type deposits.

The Central District

The Central District, about 2,000 square miles in area, lies in Morgan, Miller, Camden, Moniteau, Cooper, and Cole counties. It contains many small deposits of barite, pyrite, galena, and sphalerite, with barite being more important than the others. The occurrence and distribution of the ores here are distinctly different from those of the other two lead-producing districts, and here the ores occur in the Ordovician Gasconade, Roubidoux, and Jefferson City formations. Although lead and zinc were produced here from 1830 to 1910, the deposits are too small to be important current sources. As of this time, even most of the barite has been removed. However, in the late 1960s a minor revival of production was undertaken by a company known as Circle Mines. The mined deposits in the Central District were shallow and relatively easy to find, so geologists are not willing to say the area has been mined out until there has been exploration for deeper deposits.

The Southwest District

The southwest district is part of the world-famous Tri-State Lead-Zinc District, which extends westward into Kansas and southwestward into Oklahoma. When the ore here was first mined, the zinc ores were discarded, but in the Missouri area the zinc later became more important (see the section below on zinc). For lead, the most productive areas in the district were the Ash Grove-Everton area northwest of Springfield, the Springfield area, northern Christian County, eastern Ozark County, and the Seymour-Mansfield area in Webster and Wright counties.

The main ore minerals of the district were galena and sphalerite, and they occurred in the Mississippian rocks above the Chattanooga Shale. At first these beds were referred to as the Boone Formation, but now they are recognized as the Keokuk, Reeds Spring, and Grand Falls formations.

Galena, a combination of lead and sulfur, is the predominant lead-bearing mineral in all the lead-producing districts in Missouri. At some of the smelters, sulfuric acid is manufactured as a by-product. The importance of galena to the economy of Missouri was officially recognized in 1967 by the 74th General Assembly, which designated it as the Missouri state mineral.

Production of lead in the state has been continuous since the late 1700s. It reached an early peak in 1917 with 233,564 short tons. In 1980 the production was 548,030 short tons. Because the production of lead is highly sensitive to price fluctuations on the world metal market, production slumped severely in the 1980s. In 1986 it was 352,630 short tons, and in 1990 it was 384,976 short tons valued at $390,414,000. In 1984 seven Missouri mines produced 86 percent of the total domestic lead production. In 1990, Missouri ranked first in the nation, producing 94 percent of the United States mine output.

The average person does not see lead being used or consumed as one does iron or silver. The largest use of lead in the United States is in storage batteries, which in 1984 accounted for 72 percent of the total consumption. The average need is about 20.6 pounds per battery. Recent growth in the production of electronic devices has increased the demand for lead in radiation shielding and in the making of cathode-ray tubes in video display terminals and television picture tubes, which require lead oxide in order to improve the clarity of the glass. Other uses are in ammunition, bearings, casting metals, plumbing, solder, and gasoline additives. The latter use is being greatly reduced by federal regulation to decrease lead emissions in automobile exhaust.

Spent batteries are an important source of recycled lead metal. To take advantage of both primary and secondary lead resources, the Doe Run Company has built a battery recycling plant near its lead smelter and mines in the Viburnum Trend. When in full operation, the recycling plant will recover not only lead metal but the old battery casings as well, thereby greatly reducing the waste that would normally have gone to landfills.

Lime

The word *lime* is often used loosely to refer to limestone or lime rock, and there is often confusion between the terms *quicklime* and *hydrated lime*. When limestone or dolomite is heated in a kiln to 2,000 to 5,000 degrees F, the carbon dioxide is driven off and the resultant residue is termed *quicklime*. It may be calcium oxide or calci-magnesium oxide. If water is added it becomes *hydrated lime*.

Lime has been produced since earliest historic time and has an enormous variety of uses in chemical and industrial processes. About 50 percent is used in glassmaking, paper manufacturing, sugar refining, water softening, cement making, or as a paint filler. About 40 percent is used as a flux in metallurgical processing. The rest is used in agricultural and construction processes.

In Missouri the abundance of limestone and dolomite has provided plenty of raw material for lime production. By 1880, forty plants were operating in twenty-one counties, but most of the production has come from Ordovician beds in St. Francois and Ste. Genevieve counties and from Mississippian beds in Greene and Marion counties. The first reliable records are from 1898, when 187,000 tons were produced. By the mid–1960s production had reached 1.4 million tons valued at more than $18 million. In the last several decades the state has ranked among the top three or four producers in the nation. Most of Missouri's lime has gone to chemical and metallurgical purposes, with over 80 percent exported to other states. Production in 1987 was 1,640,000 short tons valued at $84,862,000. Reserves are abundant and will supply the need for any anticipated expansion.

Oil and Gas

Missouri has never been among the major petroleum-producing states but has had limited production of oil and gas since the late 1800s. Oil and gas are formed by the decay of organic plant or animal material that settles on the sea floor and becomes buried by sediment. The beds where this happens are called "source beds." From the source, the oil and gas migrate upward through and into porous sedimentary rocks until trapped by a nonporous confining layer. Among the simplest of these traps are anticlines, but other geologic situations create traps as well. Some of them are faults, unconformities, or natural changes in the porosity such as "pinchouts." While geologists refer to "oil pools," oil and gas do not occur in large cavities or caverns. Instead they exist in the pore spaces or small crevices in the host rock. In locating oil, the geologist is actually looking for the geologic structures that trap it, rather than for the oil itself (Fig. 7–5).

Early drilling for oil in Missouri was stimulated by the occurrence of a few surface seeps and the existence of asphaltic sands in some of the state's Pennsylvanian rocks. Gas and "oil showings" were found in a few early shallow wells, but none produced worthwhile quantities of oil. Many shallow and some deeper holes were "dry." This was discouraging because the Pennsylvanian beds contained coal and black carbonaceous shales that were known to be commonly associated with oil and gas in other states. A few wells were deepened to reach the Devonian beds that had produced oil in Oklahoma; one well in Holt County has produced oil from the Ordovician, but not in economical quantities.

The first wells were drilled in the late 1860s in Jackson County, and drilling has continued to the present. Discoveries of oil in Bates County in 1883 and in Ray County in 1886, and the finding of gas in Jackson County in 1887, stimulated more drilling. In the 1890s and early 1900s gas was discovered in Cass County and in southwestern Jackson County. Also in the first decade of the 1900s gas was found in Clay County (Fig. 7–6).

From the late 1800s to the mid–1900s, gas production was more important than oil (Figs. 7–7 and 7–8). Oil production did not exceed 40,000 barrels per year until 1954, when the Florissant Field in northern St. Louis County was discovered. In that year, production reached 96,021 barrels. Because of the low price of oil, production did not increase until the early 1980s. Annual gas production exceeded 103 million cubic feet from 1962 to 1969 but has dropped considerably since then. Since 1978 natural gas production has ranged between 4 and 5 million cubic feet. Since 1980 oil production has ranged between 100,000 and 270,000 barrels.

Most of Missouri's hydrocarbon production has been from Lower Pennsylvanian rocks, but other parts of the column have been productive. The Florissant Field and the Runamuck Field in northwestern

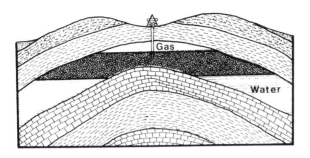

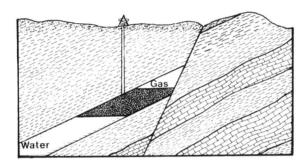

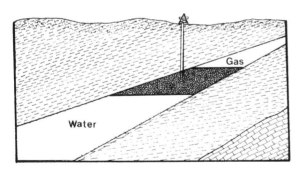

Fig. 7–5. Typical oil producing structures. The dark areas represent the zones where the rock pores are saturated with oil. Horizontal ruling indicates rock saturated with water. Clear areas denote presence of gas in rock pores.

Missouri produce from the Ordovician Kimmswick Formation. The Runamuck Field is currently the state's largest producer. From June 1987 through March 1988 it produced 33,830 barrels.

Bruce W. Netzler (1982) has suggested that the Forest City Basin of northwestern Missouri holds the possibility of good production, even though the area has been virtually unexplored. The Bootheel of Missouri shows some promise of deep production but is also largely unex-

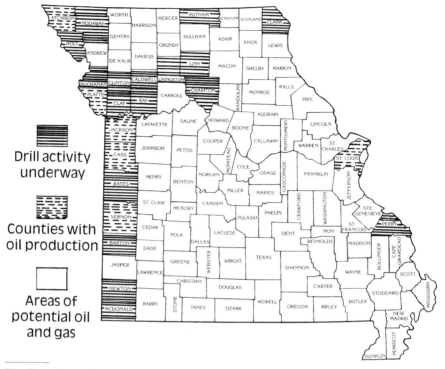

Drill activity
underway

Counties with
oil production

Areas of
potential oil
and gas

Fig. 7–6. Areas of oil and gas production in Missouri, 1978–1982.

plored. In 1988 a major oil company drilled a 10,500-foot hole in Dunklin County, hoping for a significant strike. Unfortunately, it was a dry hole, but it did set a new depth record for drilling in the state.

Asphaltic sands have been quarried from the Cherokee strata in west-central Missouri, but because of the high gravity rating and the viscosity of the oil, drilling has only been profitable when the price of oil was high enough to support secondary recovery, which involves pumping steam or specially formulated fluids into the producing formation. This procedure liquefies the viscous oil so it can be pumped from wells. Netzler (1990) has noted four asphaltic sandstones of the Cherokee that offer the potential for production of "heavy oil" from shallow wells of 100 to 150 feet using existing recovery techniques. The best prospects are in Vernon and Barton counties.

Of course the question often asked is, Why isn't there more oil in Missouri? The rocks of western Missouri do contain many of the structures that have produced oil and gas elsewhere. However, it seems that the conditions required for production of quantities and qualities of oil

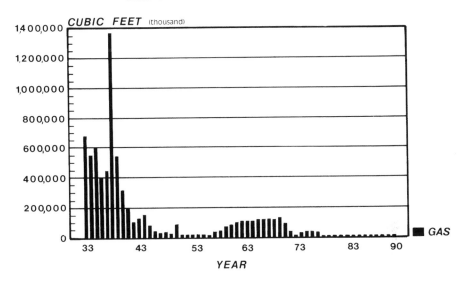

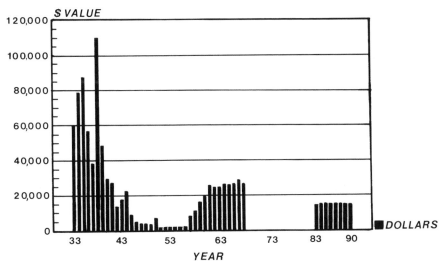

Fig. 7–7. Missouri gas production in thousands of cubic feet and in dollars, 1933–1990.

significant enough to be removed just never existed here. If they ever did, it has been suggested that the petroleum products may have been flushed away during the long period from Pennsylvanian to Cenozoic time. Research is still being done to devise methods of "enhanced oil

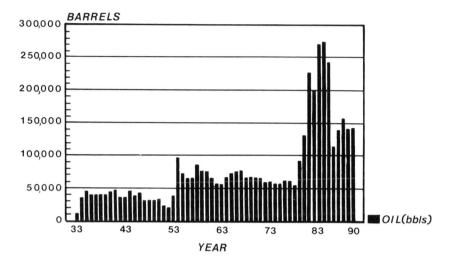

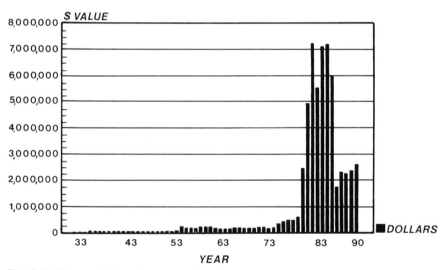

Fig. 7–8. Missouri oil production in barrels and in dollars, 1933–1990.

recovery" that might make economic production possible in areas where some oil does exist but is not recoverable by present methods.

Sand and Gravel

Because sand and gravel are so common and so familiar to everyone, it sometimes comes as a surprise to find that the total production in tons exceeds that of any other mineral commodity and that, in value, sand and gravel rank about seventh among Missouri's mineral products. The principal sources of Missouri sand and gravel are the in-channel and floodplain deposits of rivers and streams. In northwestern Missouri there is some production from the Ice Age glacial deposits. Approximately three-fourths of the sand produced comes from the in-channel deposits of the Mississippi and Missouri rivers. Most of the sand in these large streams is quartz. However, in southwestern Missouri quartz is present in smaller amounts because most of the stream sediment is chert derived from the weathering of Mississippian limestones.

The earliest year from which production figures are available is 1905, when 2 million tons valued at $668,153 were recorded. Annual production has fluctuated with economic conditions and with periods of major construction of dams and highways. In 1984 Missouri's total production of construction sand and gravel was nearly 8 million short tons valued at about $20 million. In 1990 the production was 11.8 million short tons valued at $40.2 million.

General building and paving projects use about 80 percent of the annual sand production and nearly 90 percent of the gravel, although the use of crushed limestone for concrete aggregate is reducing the need for gravel. The Mississippi, Missouri, and Meramec rivers and many Ozark streams have great potential for future supplies.

Silica Sand (Industrial Sand)

Because silica-rich minerals are hard, chemically inert, and fusible at high temperatures, modern technology requires an increasing supply in such processes as glassmaking, metal casting, grinding and polishing, and filtering. Quartz sand is a rich and readily available source of this material.

The St. Peter Sandstone from the Ordovician Period is a white, fine- to medium-grained, remarkably pure sandstone that is widely available in eastern and southeastern Missouri. C. L. Dake (1918) reported that in seventeen analyses the stone averaged 99.44 percent silica. Because of an abundance of well-rounded grains, the St. Peter is also an important "frac sand." In oil and gas fields, producing formations can be further exploited by a process called *fracturing*. Loose sand is mixed with a fluid-carrying agent and pumped into wells under very high pressure. Any cracks are enlarged and "propped" open by the sand grains. When the fluid is then removed, the sand grains hold the fracture open, increasing the permeability of the formation.

Production from the St. Peter comes from both surface and underground mining. Since 1870 the St. Peter has been quarried and mined near Crystal City in Jefferson County, and there have been operations at Festus, Pevely, Pacific, and Augusta. The St. Peter is also extensively quarried near Ottowa, Illinois. Production in 1968 exceeded 1 million tons, but in 1984 it had declined to 614,245 tons. In 1990 production was 765,000 short tons valued at $10.65 million. The remaining silica sand resources of Missouri are adequate for many more years of production.

The purity of the St. Peter Sandstone is of great importance to glassmakers. Ordinary glass is made by mixing sand with soda ash and crushed limestone and then melting the mix at temperatures around 1,100 to 1,200 degrees F. Impurities in the mix affect the color of the glass, hence purity is essential for a clear, colorless production. When colored glass is desired, coloring agents are added in the proper proportions. Iron oxide will produce a green glass (like the old Coke bottles). Blue can be obtained by adding cobalt, and pink by using selenium. For the clear crystal of fine vases and tableware, lead oxide is added with the soda ash instead of limestone.

Some industrial silica has been produced from old chat piles and mine tailings dumps near Joplin in the Tri-State Zinc-Lead District. (*Chat* and *tailings* are waste products of mining, being what is left of the mined rock after the ore minerals are removed.) In the Tri-State District, this waste material contains a high percentage of chert, which is mostly silica.

Silver

Silver is familiar to everyone for its use in coins, jewelry, and flatware. Not so well known is its use in the making of photographic film and in the making of various chemicals, astringents, and antiseptics. Missouri has never been an important silver producer, despite a rich folklore of lost silver mines. Only one mine was ever operated principally for silver, the Einstein Mine in Madison County. The source was two quartz veins in Precambrian granite, but the thinness of the veins and the difficulty of mining caused the effort to be discontinued.

The state's annual production has been generally less than a few tenths of 1 percent of the national figure. All of Missouri's silver has been a by-product of the purification of lead and zinc ores in which silver occurs as a trace element, having come into the rocks with the same mineralizing solution as the lead. Hence the production of silver has varied over the years with the price and demand for the other metals. The first year on record is 1905, when 12,900 troy ounces were produced. During some years there has been no production, but in 1952, 517,432 troy ounces were produced. Production greatly increased to 2,241,159 troy ounces in 1982 but dropped to 1,401,070 troy ounces in 1984. In 1990 Missouri produced 43 metric tons of silver valued at $6,912,000.

Stone

The words *rock* and *stone* are sometimes thought to be synonyms, but technically they are not. Stone is more properly used to refer to rock that is cut and used for a particular purpose. This may include either what we call *dimension stone*—stone trimmed to a particular size and shape for use in building—or *crushed stone*, which is used where shape and precise size are not so important.

Dimension stone of course has a higher unit value than crushed stone because it must meet strict requirements on quality, texture, and color. Most widely used are limestone and dolomite, but granite, sandstone, and slate are also commonly used. Missouri produces all of these except slate. Among the important dimension stone formations in Missouri are the Ordovician Kimmswick and Jefferson City dolomites and the Mississippian Burlington, Warsaw, and Keokuk limestones. The Graniteville granite and the Roubidoux and Gunter sandstones are also used.

Crushed and broken stone has a wide variety of uses: as road stone, cement, lime, agstone, riprap, a mineral filler, terrazzo chips, smelting flux, roofing granules, and an aggregate in concrete. Crushed and broken stone makes up about 99 percent of the total tonnage of Missouri's stone production and accounts for 17 percent of the total nonfuel mineral value. In 1990 production was 54.7 million short tons valued at $190,900,000.

Stone is a major mineral commodity and is produced in every state. In Missouri it is produced from all the Paleozoic systems and from the Precambrian, and enough reserves are available to supply all the state's need for many years to come. Because each type of stone has different uses, the production of limestone and dolomite, sandstone, granite and felsite, and marble will each be discussed in turn.

Limestone and dolomite

These two rock types are treated together because in some uses they are considered to be interchangeable. They cover about 60 percent of the state. The limestone is predominantly calcium carbonate and the dolomite is magnesium carbonate. Some formations contain both. These carbonates are the most important rocks in modern economy and for several years have accounted for 95 percent of Missouri's total stone production. Both are used as dimension stone and as crushed and broken stone.

Limestone and dolomite have been quarried in Missouri since the early settlers arrived. Many old houses still standing are built entirely of stone, and many have stone foundations and chimneys. In many areas where transportation costs are not prohibitive—for example, around Springfield and Carthage and in the Lake of the Ozarks region—stone houses are still being built. The gray limestone buildings in the eastern part of the University of Missouri campus in Columbia are of Burlington limestone quarried nearby, as are the famous columns on Francis Quadrangle. In the Kansas City area Pennsylvanian limestones are used.

Sandstone (other than silica sand)

Some sandstones in Missouri are suitable for use as dimension stone and since early time have been used for houses, patios, foundations and retaining walls, and bridge abutments and piers (for example, in Eads Bridge over the Mississippi River at St. Louis), although beneath their high-water lines piers are faced with Missouri granite. Many homes in central Missouri are veneered with colorful and ripple-marked slabs from the Roubidoux Formation and from the Gunter Sandstone Member of the Gasconade Dolomite. Asphaltic sandstone from Pennsylvanian time in Barton and Vernon counties is crushed and mixed with additional asphalt and used as road-paving material.

Granite and felsite

Granite and felsite are two of the most common rock types of the St. Francois Mountains. Quarrying of granite began about 1870, and during the early part of the twentieth century, many tons were quarried and shaped for use as curbing and paving blocks in St. Louis. The granites are both red and gray, and the red granite has been shipped all over the United States for use as monuments, construction blocks, and decorative veneers. Production has varied greatly, reaching a high in 1935 and 1950 but falling to a low in 1945.[4] Elephant Rocks State Park is an excellent place to see masses of the granite, and a granite quarry operates nearby (Fig. 7–9).

Felsite has been quarried and crushed for use as granules on asphalt

Fig. 7–9. Hayward Granite quarry near Graniteville, Missouri. DNR–Vineyard.

roofing and as concrete aggregate and road surfacing. Most of the granite and felsite production has been in Iron County.

Marble

The word *marble* means different things to different people. Strictly speaking, it refers to a metamorphic rock formed by recrystallization of a carbonate that was a limestone or dolomite before metamorphism. To the layman, sculptor, architect, or builder, marble may also be a limestone or dolomite that is amenable to being cut and polished for use as an ornamental or architectural stone. Several limestone and dolomite formations in Missouri have the quality and characteristics to be called marble in the commercial sense. Norman Hinchey (1946) made an extensive study of Missouri marble.

Marble has been produced and marketed in Missouri since about 1910. Apparently the first commercial production came in 1913 when a plant began operation near Carthage to furnish interior and exterior stone for the State Capitol Building in Jefferson City.

As with all stones, uses of marble are placed in two categories, dimension stone and crushed marble. Dimension marble refers to blocks or slabs that are quarried and cut to specific shapes for use as building blocks, veneer facings, floors, walls, stairs, wainscoting, and many other elements in a building. Dimension use requires rock without cracks, joints, or other flaws. For a time, increasing use of cast concrete reduced the demand for this type of marble, but this trend is now being reversed.

Crushed marble is used principally for terrazzo chips, exposed aggregate bits, and as patio rock for landscaping and flower gardening; it is also used for sewage filtration, ice control, and concrete aggregate. When finely ground, some is used in poultry grit, stock feed, paint, cosmetics, pharmaceuticals, and glassmaking.

Marble has been produced in Missouri from Ordovician, Devonian, Mississippian, and Pennsylvanian formations and has been shipped to all parts of the United States for use in many public buildings, only a sampling of which are mentioned in the more detailed listing below of the specific marbles from each period. For a few years in the early 1900s, cave onyx was quarried from numerous caves in central Missouri and used to make polished slabs of colorful marble for interior stonework in buildings. Fortunately for the beauty of the caves, onyx mining was short-lived because of the variability in quality of the stone.

Ordovician. In Ste. Genevieve County the Kimmswick Formation was first quarried in 1922 and was used mostly as a decorative interior stone. Five distinct colors and varieties were produced and given trade names. *Adorado* or *Ste. Genevieve Botticino* is a white, light gray, or creamy white marble that has been used in the construction of the National Gallery of Art and the Library of Congress in Washington, D.C., and in post office buildings in Tulsa, Newark, and Des Plaines, Illinois. *Shelldorado* is a light gray marble mottled with white or dark gray fossils; it was used for the post office in Enid, Oklahoma. *Eldorado* or *Ste. Genevieve Istrian* marble, which is gray to brownish gray and fossiliferous, with pinkish areas and stylolites or veins of "gold color," has been used in post offices in Tillamook, Oregon; Red Bud, Nebraska; Traverse City, Michigan; and Ste. Genevieve, Missouri; in high schools in Philadelphia and Montpelier, Vermont; and in the city hall of Kansas City, Missouri. *Indorado* or *Inkley Vein* marble is dark gray streaked with bluish gray and yellowish brown with sections of white where it contains fossils; it has been used in the Dallas airport terminal. Finally, *Gradorado* is brownish gray (lighter than Indorado) and shows small white fossil sections; it was used in the Hall of Music at Indiana University in Bloomington.

Devonian. Also in Ste. Genevieve County the Devonian Grand Tower and Little Saline formations have been important sources of marble. *Ste. Genevieve Rose* is the trade name for a marble with variegated colorings chiefly of rose, pink, deep red, and greenish gray. It has been used in Kiel Auditorium and Soldiers' Memorial in St. Louis; for the courthouse in Austin, Texas; and in post offices in Miami; Council Bluffs, Iowa; Jackson, Mississippi; St. Louis; Rochester, New York; Columbus, Ohio; and Portland, Oregon. *Ste. Genevieve Gold Vein* has a light to medium gray body with delicate veining in a buff to tan, light and dark coloring. It shows cross sections of corals and crinoid fragments and has been used in the same buildings as the Ste. Genevieve Rose, as well as in the Los Angeles city hall; the Montreal courthouse; the Department of Commerce and National Archives buildings in Washington, D.C.; the Bank of Hawaii in Honolulu; the Chicago Hilton; Abraham Lincoln's tomb in Springfield, Illinois; the American Museum and Rockefeller Center in New York City; the naval hospital in Philadelphia; and the Cathedral of St. Louis.

Mississippian. The Warsaw and the Burlington-Keokuk beds in Jasper and Greene counties are also well known and widely used for interior decorative marble. The rock quarried in the old Phenix Marble quarry in Greene County is from the Burlington-Keokuk Formation. This stone is known as *Napoleon Gray* marble when sawed across the bedding and as *Anellen* when sawed with the bedding. Both are dark to slightly brownish gray and take a high polish. Representative installations are the Los Angeles city hall; the Missouri State Capitol; the Nelson Gallery–Atkins Museum, the Post Office Building, and the Federal Courts Building in Kansas City; the Civil Courts Building in St. Louis; the National Academy of Sciences in Washington, D.C.; and the New York Stock Exchange.

The Warsaw Formation is quarried northwest of Carthage and is generally known as Carthage Marble. As with the Phenix beds, the trade name varies with the direction of sawing or the method of finishing. Cut blocks finished by sand rubbing are called *Carthage Exterior* and used for outside installations. The Buehler Building in Rolla, headquarters for the Division of Geology and Land Survey, and the Capitol in

Jefferson City are of Carthage Marble. *Ozark Veined* is sawed across the bed and is used as an interior marble in the U.S. Federal Court Building in New York City and on the campus of Notre Dame University in South Bend, Indiana. *Ozark Tavernelle* is sawed with the bed. It is used inside the Interstate Commerce Commission and for the fountains at the east end of the National Art Gallery in Washington, D.C., as well as inside the Rosenwald Museum of the University of Chicago and the Municipal Auditorium in Kansas City. *Ozark Fleuri* is sawed with the bed but has clouded dark areas. Representative installations are on the campus of Stanford University in Palo Alto, California, and inside the Federal Hospital in Springfield, Missouri. *Carthage Interior* is used inside the State Capitol Building in Jefferson City and the post office buildings in St. Louis, Jefferson City, Springfield, and Joplin.

Pennsylvanian. Although the Pennsylvanian has several limestone formations, only one location has produced polished stone for building use. Norman Hinchey (1946) reported that excavation for the basement of the Nelson Gallery–Atkins Museum in Kansas City encountered two limestones that would take a high polish. The Argentine Member of the Wyandotte Formation was used for the walls of the vestibule, and the Raytown Member of the Iola Formation was used at the base of the walls. Because it contains fossils that show up as light-colored streaks on polished surfaces, the Raytown has become known as *Calico Rock*.

Because of the widespread use of Missouri marbles in buildings in major cities across the country, one can truly say, "There's a bit of Missouri in every major city in the United States."

Tripoli

Tripoli is a very fine-grained siliceous material used primarily as an abrasive or polishing powder. It mainly occurs as a residue after intense chemical weathering of highly siliceous limestone. Although it is no longer being produced in Missouri, tripoli is mentioned here because it was at one time mined in southwestern Missouri, near Seneca and Racine in Newton County, and until 1914 Missouri was the leading producer in the United States. The Missouri deposits were derived from limestone in what is now called the Warsaw Formation. Production values cannot be obtained because the U.S. Bureau of Mines lumps figures from the Missouri-Oklahoma district to avoid disclosing information about individual companies. However, the values are small.

Tungsten

The most familiar use of this metal is in the filaments of electric light bulbs. But another very important one is as an alloy in steel made for high-temperature use, or where very hard cutting tools are required. Tungsten has never been significant in Missouri's mineral production, accounting for only a fraction of 1 percent of the United States total. The only production was from the Einstein Silver Mine in Madison County, where tungsten occurs in Precambrian granite and felsite.

Water

In contrast with many states, Missouri is favored with a water supply that is one of its most important natural resources. With the Mississippi forming the state's eastern boundary and the Missouri forming its north-western boundary and then flowing eastward across its middle, much of the state has abundant supplies of surface water—sometimes too much, as the floodplains of these rivers demonstrate. The many tributaries of these major streams also contribute to the surface supplies. In the rugged topography of the southern part of the state, many streams have been impounded, forming the Lake of the Ozarks, Stockton Lake, Table Rock Lake, Harry S. Truman Reservoir, and Pomme de Terre Lake.

To supply these streams, the annual average rainfall in Missouri ranges from about 32 inches in the northwest to 48 inches in the extreme southeast. Unfortunately, the rainfall varies greatly from year to year, causing both floods and droughts. Normally the heaviest rainfall occurs between May and August, and in some years torrential summer storms cause extensive damage. A storm at Holt in 1947 dropped twelve inches of rain in forty-two minutes. This is believed to be the most intense rainfall ever recorded anywhere in the world.

Many of the streams in southern Missouri are fed by springs. Although their flow is ultimately derived from rainfall, there is a delay in the response because the water must filter through the rocks feeding the springs. These streams have well-sustained base flows and therefore are less likely to experience flash floods or frequent large fluctuations. Many of these spring-fed streams are popular for float trips, trout fishing, camping, and other recreation. Some are used for commercial fish propagation as well as for sport fishing.

The Mississippi and Missouri rivers are also important as avenues of water-borne commerce. Barges pushed by diesel towboats move Mid-

western grain to markets, distribute Missouri cement to far-flung regions, and carry coal to power plants.

River flow is used to generate power at seven sites. The Osage is used
at two sites, Truman and Bagnell dams; the White also at two, Ozark
Beach (Powersite Dam) and Table Rock; and the Niangua at Tunnel
Dam. Tunnel Dam and Powersite Dam are minor contributors to the
power supply, but Bagnell provides 43.4 percent of the hydroelectric
power of the state and Table Rock 48.7 percent. In northeastern Missouri, the recently completed Cannon Dam on the Salt River also has
power plants. The seventh site is along the Black River, which is fed by
many springs and provides water for a rather unusual hydroelectric
project. On the East Fork of the Black at Taum Sauk, the Union Electric Company built an impoundment from which water is pumped to a
higher reservoir on Proffit Mountain during periods of low demand, to
be released to generators during periods of peak demand.[5]

As sources for municipal or domestic supplies, the surface streams
require filtration and chemical treatment. Fortunately, Missouri also
has a large supply of ground water available through properly con-
structed wells. The four main sources for ground water are the Cambrian and Ordovician formations of the Ozark Province, the Cretaceous McNairy Formation of the Southeastern Lowlands, the Wilcox
Group of the same area, and the alluvial fill in the valleys of the Southeastern Lowlands and along the major rivers (see Plate 6).

The Ozark area, excluding the St. Francois Mountains but including
the Springfield Plateau, has the most extensively used fresh ground-
water supplies. Here, more than 2,000 feet of Cambro-Ordovician water-
bearing limestones, dolomites, and sandstones are overlain by Mississippian limestone. Because the carbonate rocks are soluble and therefore
porous and cavernous, the water is hard, and there is great potential for
pollution and contamination. This threat requires careful drilling, ce-
menting, and sealing of the wells. However, large quantities of ground
water can be obtained almost anywhere in this Ozark province, and wells
yielding more than 1,000 gallons per minute are common.

Unfortunately, in some areas the ground waters are saline and too
highly mineralized for general use. This occurs mostly in areas overlain
by Pennsylvanian shales, which prevent the downward movement of
rainfall that would otherwise flush out the old seawater trapped in the
underlying beds. As discussed in Chapter 5, the highly mineralized waters were once popular for health spas and for medicinal uses.

The McNairy Formation in the Southeastern Lowlands is composed
mostly of poorly consolidated sandstone; it contains some clay layers that

are nearly impermeable and hence do not yield much water. However, in some locations in the Southeastern Lowlands the sandstone yields water that is soft, low in iron, and thus desirable for municipal use and for irrigation. The Wilcox Group, composed also of sand and clay layers, forms an aquifer that yields relatively soft, low-chloride water for municipal use.

In much of the area north of the Missouri River, glacial drift deposits and buried, gravel-filled, preglacial stream valleys provide substantial water supplies to relatively shallow wells. During the Ice Age, the melt water from the glaciers created many high-velocity streams. These cut deep river valleys that have since been filled with sand and gravel to considerable depths. In the major stream valleys these alluvial deposits are sources of large quantities of water. In the Missouri River valley, alluvial deposits range from 80 to 100 feet in thickness, and in the Southeastern Lowlands they exceed 200 feet. The water level is usually about 10 to 20 feet below the land surface, and yields of 2,000 to 3,000 gallons per minute are not unusual. Columbia is a city that obtains water from alluvial deposits along the Missouri River. Ground water pumped from several wells is piped northward to the city, where it is treated and made available for homes and businesses.

Zinc

Zinc is a more common metal than most people realize because much of its use is in alloys or in die-cast items where plating conceals the metal. As an alloy, zinc is an important component of brass for use in lamp fixtures, radiator cores, screws, watch and clock parts, cartridge cases, and many small parts in electrical or electronic mechanisms. Zinc's largest use is in die-cast items, because it can be melted and cast at relatively low temperatures and pressures and when cast in smooth molds produces a smooth surface that can be readily plated. In this form, zinc is found in automobile door handles, window cranks, fuel pumps, tools, and toys.

Because zinc resists normal weathering processes, it is also used as a protective coating for other metals that would deteriorate when exposed. This is especially true for iron, and the process of covering iron with zinc is called *galvanizing*. Common galvanized items are nails, metal roofing, bolts, wire mesh, fence wire, hinges, buckets, pipes, and water tanks. When made into thin sheets, zinc is used for photoengraving plates and as a casing for electric anodes and dry-cell batteries. Finally, zinc in the form of zinc oxide is used as a filler in paints, ceramics, textiles, floor coverings, and rubber tiles. It is also popular as a sun-screen to prevent sunburn.

Three districts in Missouri have produced zinc ores. The southwest district, which is part of the Tri-State Zinc-Lead District, has been the most productive. Here the ore was mined primarily for the zinc, with lead as a by-product. Most of the ore was produced from Jasper and Newton counties from the Mississippian Keokuk and Warsaw forma-tions. Production values for Missouri are not available because the data were compiled for the total district. Zinc mining in the Missouri part of the district was discontinued in 1957.

The southeast district is the only one where zinc mining continues today. Here, zinc comes as a by-product of lead mining, and as the lead market fluctuates, the zinc production will probably parallel it. The ores in this district also produce limited amounts of copper and silver as by-products. The profitable ores are mostly in the Cambrian Bonne-terre Formation with smaller amounts in the Lamotte Sandstone.

The Central Mineral District in Morgan and Miller counties is of less importance because the ore deposits are smaller and are scattered. Zinc production here stopped in 1945. The producing beds were in the Rou-bidoux and Jefferson City formations.

Prior to 1918 Missouri led all states in the production of zinc ores. A peak was reached in 1915. During this time, the yearly total was more than 100,000 short tons, and the annual dollar value (except for 1900 and 1901) exceeded $10 million. In 1990, production was 51,555 metric tons valued at $85,244,000, and Missouri ranked third among the states.

EPILOGUE

And some rin up hill and down
   dale, knapping the chucky
   stanes to pieces wi' hammers,
   like sae many road makers run daft.
They say it is to see how the warld was made.
     —Sir Walter Scott, "St. Ronan's Well," 1824

Geologists have never quit "knapping the chucky stanes to pieces." We have even extended the "knapping" to the moon and to Mars. Besides wanting to see how the world was made, we want to see what it is made of, and how we can use it.

Over the years the science of geology has developed many branches, and geologists, like the members of most professions, have specialized. It is easy to understand that the first branch was for the finding and recovery of metal ores, so that the mining geologist came on the scene early. Today there are oil geologists, coal geologists, hydrogeologists, structural geologists, seismologists, and many others. We have found ways to exploit all of the earth's resources, and to understand most of its processes. We put these resources to work to make our lives more comfortable and productive.

However, about the middle of the twentieth century we began to realize that many of our activities are maliciously abusing our precious resources. We know that we are consuming resources that have finite limits, and that they are not being replaced. Coal and oil have taken millions of years to form, and no new deposits are coming into being— at least none that would be available to us. Fresh ground water in storage for thousands of years is being pumped from extensive aquifers and not being replaced. We also now realize, belatedly, that our activities are polluting our atmosphere and our surface water supplies. Lakes and rivers have become hazardous "chemical soups," and we are learning how expensive cleanups can be. We are burying our trash and solid waste to get it out of sight and out of mind, but in doing so we are also polluting and contaminating our out-of-sight mineral resources.

We have found many other problems that will need to be addressed.

How can we expand our limited supply of building material? How can we develop new and more efficient fuels? How can we better use our land? How can we be sure we are developing appropriate building sites, better highways, safer airports? How can we control floods and landslides? How can we learn to predict or even prevent earthquakes? The solutions to these problems will affect each of us, and our understanding and management of geologic resources will play a large part. Our "environmental scientists" will need our "environmental geologists," who will also need to call on all other geologists, so that we can use the vast volume of geologic knowledge to help protect us from ourselves.

NOTES

Chapter 1: ROCKS OF MISSOURI

1. The ages reported here are from a U.S. Geological Survey report, listed in the Bibliography as Marvin, 1988.

2. In modern oceans, there is widespread primary deposition of limestone and limy muds, so it is not difficult to account for the great volume of limestone in Missouri. However, there are no widespread areas of primary deposition of dolomite. Hence the great thicknesses of dolomite must once have been limestone, which was converted to dolomite by the introduction of magnesium-rich ground water.

3. See also the works by Keller, Kisvarsanyi, Hinchey, and Tolman and Robertson listed in the Bibliography.

Chapter 4: STRUCTURAL FEATURES

1. For an excellent nontechnical history of the earthquake see Penick, 1981. For comprehensive scientific data see Nuttli, 1973; McKeown and Pakiser et al., 1982; and Johnston, 1982. Also see Hamilton and Johnston, 1990.

2. W. P. Strickland, ed., *Autobiography of Rev. James B. Finley; or, Pioneer Life in the West* (1854).

3. The report by Drake was in his *Natural and Statistical View or Picture of Cincinnati and the Miami Country* (1815).

4. See McKeown and Pakiser et al., 1982.

Chapter 5: KARST

1. *National Speleological Society News* (September 1989).

2. See Bretz, 1956, pp. 151, 198, and pl. 167.

3. Weaver and Johnson, 1980.

4. Extensive reports on Missouri caves are available in Bretz, 1956, and Weaver and Johnson, 1980. Color photographs of caves and an assortment of speleothems can be found in Weaver, Huckins, and Walk, 1992. In addition, the Missouri Speleological Survey has published a cave science journal called *Missouri Speleology* since 1969, and the library at Central Missouri State University in Warrensburg maintains a collection of speleological literature.

5. U.S. Senate, 1967. The one-billion-gallon flow was observed during a heavy flood that destroyed all the measuring devices.

6. See Schweitzer, 1892.

7. Vineyard and Feder (1982) illustrate many examples of underground streams.

8. Reports by Beveridge (1978) and by Vineyard and Williams (1965) illustrate and document much sinkhole activity in Missouri.

Chapter 6: THE MISSOURI COLUMN

1. The entire system is described in Howe, 1961; currently being revised by Thompson, 1986, et seq.

2. The columns used here are modified from Howe, 1961.

3. One of the most detailed is that of Tolman and Robertson, 1969.

4. All these features are located on maps and described in Beveridge's *Geologic Wonders and Curiosities of Missouri* (1978) and the revised edition edited by Vineyard (1990).

5. For a helpful guide to the Precambrian rocks of the St. Francois Mountain area, see Kisvarsanyi, 1976.

6. See Howe, 1961, for a complete discussion of the Devonian rocks of Missouri by John W. Koenig.

7. For additional representations of the Mississippian succession throughout the state, see Howe, 1961.

8. For additional information, see Howe, 1961.

9. The detailed characteristics of these widespread blankets of material left by the melting ice were more fully described by Heim, 1961, using the then-accepted terms *Nebraskan* and *Kansan*. However, more recent research in Illinois, Iowa, and Nebraska indicates no basis for direct correlation of the tills in Missouri and Kansas with the better-defined sequences in Iowa and Nebraska. So it appears now that the Missouri till is better referred to as *pre-Illinoian* and that the archaic terms *Kansan, Aftonian,* and *Nebraskan* should be abandoned as stratigraphic terms (Hallberg, 1986).

10. Several erratics have been described by Beveridge, 1978.

11. See Weaver and Johnson, 1980; Mehl, 1962; and Parmalee, Oesch, and Guilday, 1969.

12. See Holst, 1988, and Merrill, 1924.

Chapter 7: MINERAL AND GEOLOGIC RESOURCES

1. For more information on mineral production and values in Missouri, see Wharton et al., 1969.

2. Greater elaboration on the history, occurrence, varieties, and uses of various types of clay is presented in U.S. Senate, 1967.

3. See Wharton et al., 1969.

4. See Wharton et al., 1969, for statistics.

5. There are numerous potential but undeveloped locations where hydroelectric plants could be built. For descriptions and more details regarding these, see U.S. Senate, 1967.

GLOSSARY

Agstone: Rock quarried, crushed, and spread on farm fields to condition the soil; most commonly limestone and dolomite.

Algae: Simple, one-celled photosynthetic aquatic plants including seaweeds and their freshwater relatives.

Alluvium: Sediment deposited in stream beds, on flood plains, or in deltas.

Andesite: A fine-grained, dark-colored extrusive igneous rock.

Anticline (Convex): An upward fold in layered rocks.

Aquifer: A bed or porous rock formation that will hold water in storage and that will yield water to a well.

Arenaceous: Containing significant amounts of sand grains.

Argillaceous: Containing significant amounts of clay minerals.

Artesian: Used to describe ground water that is under sufficient pressure to rise above the level of a confining layer.

Basalt: Igneous rock that is dark colored and fine grained.

Basic: Used to describe an igneous rock that is comparatively low in silica content.

Batholith: A large mass of intrusive igneous rock that cooled and crystallized at a depth of hundreds of feet.

Bed: An individual layer of a rock mass consisting of layers of varying thickness and characteristics, commonly referred to as being bedded. Geologists thus refer to coal beds, limestone beds, and so forth. The nature of the bedding may be a distinctive characteristic of a formation, for example, even bedded, cross bedded, thickly bedded.

Bentonite: Altered volcanic ash. That in the Ordovician beds is believed to have come from volcanic events southwest of the present Tennessee-Georgia state line.

Brachiopod: Marine shelled invertebrate animal with two equal but laterally symmetrical valves.

Bryozoa: Small colonial marine animals that build a calcareous skeletal structural support. Colonies commonly resemble mossy growths.

Calcareous: Containing calcium carbonate; limy.

Carbonaceous: Used to describe rocks or minerals whose composition is mostly of carbon, such as coal.

Carbonate: Used to describe rocks at least half of whose composition consists of minerals containing the carbonate radical CO_3. The most common ones are limestone and dolomite.

Carboniferous: As used by geologists, a time term designating the Pennsylvanian and Mississippian periods.

Cephalopod: A marine invertebrate mollusk related to the squid.

Charcoal iron: Iron separated from impurities in the ore through a process using a furnace fired by charcoal.

Clastic: Pertaining to a rock or sediment composed principally of broken fragments.

Colonial: Used to describe a grouping of individuals to form a common skeletal mass. Often occurs among corals and bryozoa.

Concretion: A nodular, spherical, or oblate object composed of clay, chert, or other mineral matter that has accumulated around a nucleus. Normally harder than the surrounding rock beds.

Continental drift: A theory that the continental masses were once joined as a huge continent, which broke into separate continents that drifted apart and that are still moving.

Craton: A portion of the earth's crust that maintains stability for a prolonged span of time, without significant deformation.

Crinoid: A marine invertebrate animal consisting of a "cup" with "arms" and a "stem." Resembles a plant and is commonly called a "sea lily." Very abundant as fossils in Mississippian limestones in Missouri.

Cyclothem: A sequence of sedimentary beds that represents one cycle of marine invasion and retreat over a coal swamp area such as prevailed during the Pennsylvanian period.

Diabase: A fine-grained, dark igneous rock of basaltic composition.

Diatreme: A general term for a "pipe" in sedimentary rocks formed by explosion of gas from an underlying magma.

Dike: A tabular body of igneous rock that cuts across the structure of adjacent rock masses; formed by the intrusion of magma that cooled deep below the surface.

Drift: A mixture of rocks and rock fragments of varying size, from clay to boulders, deposited by melting glaciers.

Druse: A rock surface covered with an aggregate of small, well-formed crystals, usually quartz.

Epicenter: The point on the earth's surface directly above the focus of an earthquake.

Extrusive: Used to describe rocks derived from magma that poured out on or was ejected at the earth's surface.

Focus: Used to describe the point within the earth at which an earthquake begins.

Formation: A rock unit that has uniform characteristics, is tabular in form, and has thickness and areal extent. Normally, the term should be used for units that are mappable at a practical or convenient scale. Proper usage restricts the word to this meaning, but it is sometimes loosely applied to other features such as natural bridges, rock pedestals, stalactites, and so forth.

Fossil: Any evidence found naturally in the rocks that represents a former living organism of past geologic or prehistoric time.

Gastropod: A mollusk with an external shell that is asymmetrically coiled, such as a snail.

Geode: A hollow, ball-shaped body, usually of calcite or quartz and lined with crystals or with a smooth lining of the same mineral. Geodes vary in size from less than an inch to more than a foot across.

Geographic center: The center of gravity of the surface of an area, or that point on which the surface of the area would balance if it were a plane of uniform thickness.

Gondwana: Large fragment of Pangea that broke free to form South America, Africa, and Australia.

Hot spot: A location underlain by a source of magmatic heating giving rise to volcanic activity.

Igneous: Used to describe rocks formed by the cooling of molten magma.

Intrusive: Term applied to igneous rock that, while still molten, invaded another rock mass and cooled at depth.

Lithographic: Used to describe compact, dense, exceedingly fine-grained rock. Usually applied to limestone.

Lithology: General term that refers to the texture, mineral content, and other physical features of a rock formation or specimen.

Magnetite: An iron-rich mineral that is naturally magnetic.

Mercalli Scale: A means of defining the intensity of an earthquake, ranging from I (detectable only by instruments) to XII (total destruction).

Metamorphic: Name applied to rocks that have been formed by the extreme heating or compressing of previously existing rock. Meta = change, morph = form.

Mica: A silicate mineral with perfect basal cleavage forming very thin sheets. Common in granites. Sometimes called "isinglass."

Onyx: A color-banded body of quartz or calcite. May be polished and used as a gemstone.

Orogenic: Term used to describe the process of formation of mountains.

Outcrop: A place, or an area, where a formation is exposed to view or where it is known to be just under the surface.

Pangea: Original large continental mass that broke up to form present-day continents.

Pinchout: An oil trap formed where a porous bed thins laterally and lies between two nonporous, confining beds.

Porphyry: An igneous rock that contains large crystals in a groundmass of very fine texture. These large crystals are commonly called phenocrysts.

Richter Scale: A means of defining the magnitude of an earthquake, ranging from 1 (barely detectable) to 9 (total destruction).

Road metal: Crushed stone used for surfacing unpaved roads.

Sedimentary: Term applied to rocks formed of clastic fragments or chemical components of previously existing rocks.

Shut-ins: Narrow, constricted segments of a stream valley where the stream has had to cut through resistant rocks.

Syncline: A fold in layered rocks in which the beds are convex downward.

Tectonics: Used to designate the study of the movements of portions of the earth's crust.

Till: Nonsorted rock material deposited by melting glaciers.

Tuff: A consolidated rock composed of fragments of material ejected from a volcano.

Unconformity: A surface of contact between two rock units that represents a gap in deposition or a period of erosion.

Vug: A small cavity in solid rock. More common in sedimentary rocks. May be an empty void or may be lined with crystals.

Weight measurements

Ounce

ounce avoirdupois = 28.349 grams
ounce troy = 31.103 grams

Pound

pound avoirdupois = 16 ounces avoirdupois = 453 grams
pound troy = 12 ounces troy = 373 grams

Ton

short ton = 2,000 pounds avoirdupois = 907,000 grams
long ton = 2,240 pounds avoirdupois = 1,014,720 grams
metric ton = 2,204 pounds avoirdupois = 1,000,000 grams

BIBLIOGRAPHY

Anon. 1988. *Earthquakes in Missouri*. Rolla: Missouri Department of Natural Resources, Division of Geology and Land Survey. 14 pp.

Amsden, T. W. 1974. *Late Ordovician and Early Silurian Articulate Brachiopods from Oklahoma*. Oklahoma Geological Survey Bulletin, vol. 119. 154 pp.

Beveridge, Thomas R. 1978. *Geologic Wonders and Curiosities of Missouri*. Rolla: Missouri Department of Natural Resources, Division of Geology and Land Survey. Educational Series, no. 4. 460 pp., 163 figs. Revised edition, 1990, edited by Jerry D. Vineyard.

Branson, E. B. 1944. *Geology of Missouri*. Columbia: University of Missouri Studies, vol. 19, no. 3. 535 pp.

Bretz, J. Harlen. 1956. *Caves of Missouri*. Rolla: Missouri Division of Geological Survey and Water Resources. Series 2, vol. 39. 491 pp., 168 figs.

Bunker, Bill, et al. 1988. "Phanerozoic History of the Central Midcontinent." In *The Geology of North America*, 243–57. Geological Society of America Bulletin D-2.

Cargo, David, et al. 1982. *Guidebook to the Petroleum Geology of Western Missouri*. Kansas City: Association of Missouri Geologists, 29th Annual Meeting. 77p.

Dake, C. L. 1918. *The Sand and Gravel Resources of Missouri*. Missouri Bureau of Geology, 2d ser., vol. 15. 290 pp.

Fuller, Myron. 1912. *The New Madrid Earthquake*. U.S. Geological Survey Bulletin 494. Washington: U.S. Government Printing Office. 119 pp.

Haas, Kim E. 1982. "Oil and Gas in Missouri, 1978–1982." In *Guidebook to the Petroleum Geology of Western Missouri* by David Cargo et al. Kansas City: Association of Missouri Geologists, 29th Annual Meeting.

Hallberg, George R. 1986. "Pre-Wisconsin Glacial Stratigraphy of the Central Plains Region in Iowa, Nebraska, Kansas, and Missouri." *Quaternary Science Reviews* 5:11–15.

Hamilton, Robert M., and A. C. Johnston. 1990. *Tecumseh's Prophecy: Preparing for the Next New Madrid Earthquake*. U.S. Geological Circular 1066. Washington: U.S. Government Printing Office. 30 pp.

Heim, George E., Jr. 1961. "Quaternary System." In *The Stratigraphic Succession in Missouri* by Wallace B. Howe, pp. 130–36. Rolla: Missouri Geological Survey and Water Resources. Vol. 40, 2d ser.

Hinchey, Norman. 1946. *Missouri Marble*. Report of Investigations, no. 3. Rolla: Missouri Geological Survey and Water Resources. 47 pp.

Holst, Sue. 1988. "Missouri's Ice Age Mastodons." *Midwest Motorist* 59, no. 6 (July–August): 47–50.

Howe, Wallace B. 1961. *The Stratigraphic Succession in Missouri*. Rolla: Missouri Geological Survey and Water Resources. Vol. 40, 2d ser. 185 pp.

Hubbell, Sue. 1991. "Earthquake Fever." *New Yorker*, February 11, 1991.

Hutton, James. 1788. *Royal Society of Edinburgh Transactions*. Vol. 1.

Johnston, A. C. 1982. "A Major Earthquake Zone on the Mississippi." *Scientific American* 246, no. 4 (April): 60–68.

Keller, W. D. 1961. *Common Rocks and Minerals of Missouri*. Rev. ed. Columbia: University of Missouri Press. 78 pp.

————. 1979. *Diaspore: A Depleted and Non-renewable Mineral Resource of Missouri*. Educational Series, no. 6. Rolla: Missouri Department of Natural Resources, Division of Geology and Land Survey. 40 pp.

Kiilsgaard, T. H., W. C. Hayes, and A. V. Heyl. 1967. "Lead and Zinc." In *Mineral and Water Resources of Missouri*, 41–63. Rolla: Missouri Geological Survey and Water Resources. Vol. 43.

Kisvarsanyi, Eva B. 1976. *Studies in Precambrian Geology, with Guide to Selected Parts of the St. Francois Mountains*. Report of Investigation, no. 61. Rolla: Missouri Department of Natural Resources, Division of Geology and Land Survey.

Lee, Wallace. 1941. "Preliminary Report on the McLouth Oil Field." *Kansas State Geological Survey Bulletin* 38, pt. 10, pp. 261–84, pls. 1–3.

Legget, Robert F. 1973. *Cities and Geology*. New York: McGraw Hill. 579 pp.

McCracken, Mary H. 1971. *Structural Features of Missouri*. Report of Investigations, no. 49. Rolla: Missouri Geological Survey and Water Resources. 100 pp., 1 pl., 11 figs.

McKeown, F. A., and L. C. Pakiser et al. 1982. *Investigations of the New Madrid, Missouri Earthquake Region*. Washington, D.C.: U.S. Geological Survey. Professional Paper 1236. 210 pp.

Marvin, Richard F. 1988. *Radiometric Ages of the Basement Rocks in the Northern Midcontinent USA*. Washington, D.C.: U.S. Geological Survey. Miscellaneous Field Studies Map MF 1835c.

Mehl, Maurice G. 1960. "The Relationships of the Base of the Mississippian System in Missouri." *Journal Scientific Labs*, Denison University, 45:57–107.

————. 1962. *Missouri's Ice Age Animals*. Educational Series, no. 1. Rolla: Missouri Geological Survey and Water Resources. 104 pp.

Mehl, Maurice G., et al. 1966. *The Grundel Mastodon*. Report of Investigations, no. 35. Rolla: Missouri Geological Survey and Water Resources. 28 pp., 8 figs.

Merrill, George P. 1924. *The First One Hundred Years of American Geology*. New Haven: Yale University Press. Reprint. New York: Hafner Publishing Co., 1964. 773 pp.

Nelson, Paul W. 1987. *The Terrestrial Natural Communities of Missouri*. The Missouri Natural Areas Committee, Missouri Department of Natural Resources. 197 pp.

Netzler, Bruce W. 1982. "Recent Oil and Gas Activities in Missouri." In *Guidebook to the Petroleum Geology of Western Missouri* by David Cargo et al. Kansas City: Association of Missouri Geologists, 29th Annual Meeting.

————. 1990. "Heavy-oil Resources Potential of West-Central Missouri." Open-File Series, OFR 90-80-OG. Rolla: Missouri Department of Natural Resources, Division of Geology and Land Survey. 10 pp.

Nuttli, Otto W. 1973. "The Mississippi Valley Earthquakes of 1811 and 1812: Intensities, Ground Motion and Magnitudes." *Bulletin of the Seismological Society of America* 63 (February): 227–28.

Offield, T. W., and H. A. Pohn. 1979. *Geology of the Decaturville Structure, Missouri*. U.S. Geological Survey Professional Paper 1042. Washington: U.S. Government Printing Office. 48 pp.

Parmalee, Paul W., Ronald D. Oesch, and John E. Guilday. 1969. *Pleistocene and Recent Vertebrate Faunas from Crankshaft Cave, Missouri*. Springfield: Illinois State Museum, Reports of Investigations, no. 14. 37 pp.

Penick, James Lal, Jr. 1981. *The New Madrid Earthquakes*. Rev. ed. Columbia: University of Missouri Press. 176 pp.

Pratt, Walden, and Martin B. Goldhaber. 1990. *Mineral Resources Potential of the Midcontinent*. Missouri Geological Survey Symposium, St. Louis, Program and Abstracts. Washington, D.C.: U.S. Geological Survey. Circular 1043.

Robertson, Charles E. 1973. *Mineable Coal Reserves of Missouri*. Report of Investigations, no. 54. Rolla: Missouri Geological Survey and Water Resources. 71 pp., 20 figs.

Ross, Rueben J., Jr., et al. 1982. *The Ordovician System in United States*. International Union of Geological Sciences, Publication no. 12.

Schweitzer, Paul. 1892. *Mineral Waters of Missouri*. Rolla: Geological Survey of Missouri. Vol. 3. 256 pp., 33 pls., 11 figs.

Searight, Walter V. 1967. "Coal." In *Mineral and Water Resources of Missouri*, 235–52. Rolla: Missouri Geological Survey and Water Resources. Vol. 43.

Searight, Walter, and Wallace Howe. 1961. "Pennsylvanian System." In *The Stratigraphic Succession in Missouri* by Wallace B. Howe, pp. 78–121. Rolla: Missouri Geological Survey and Water Resources. Vol. 40, 2d ser.

Sheehan, P. M. 1973. "The Relation of Late Ordovician Glaciation to the Ordovician-Silurian Changeover in North American Brachiopod Faunas." *Lethaia* 6, no. 2:147–54.

Sloss, L. L. 1963. "Sequences in the Cratonic Interior of North America." *Geological Society of America Bulletin* 74, no. 2 (February): 93–114.

———. 1982. *The Midcontinent Province: United States*. Boulder: Geological Society of America. DNAG Special Publication 1, pp. 27–39.

———. 1988. "Tectonic Evolution of the Craton in Phanerozoic Time." In *The Geology of North America*, 25–51. Geological Society of America Bulletin D–2.

Snyder, Frank, and Paul Gerdeman. 1965. "Explosive Igneous Activity along an Illinois-Missouri-Kansas Axis." *American Journal of Science* 263, no. 6 (June): 465–93.

Snyder, Frank, and James Williams et al. 1965. *Cryptoexplosive Structures in Missouri*. Report of Investigations, no. 30. Rolla: Missouri Geological Survey and Water Resources. 71 pp.

Spreng, A. C. 1961. "The Mississippian System." In *The Stratigraphic Succession in Missouri* by Wallace B. Howe, pp. 49–78. Rolla: Missouri Geological Survey and Water Resources. Vol. 40, 2d ser.

Steyermark, J. A. 1941. *Phanerogamic Flora of the Fresh-water Springs in the Ozarks of Missouri*. Chicago: Field Museum of Natural History. Botanical Series, vol. 9, no. 6. 618 pp.

Templeton, J. S., and H. B. Willman. 1963. *The Champlainian Series in Illinois*. Urbana: Illinois State Geological Survey. Bulletin 89. 260 pp., 41 figs.

Thompson, Thomas L. 1972. *Conodont Biostratigraphy of Chesterian Strata in Southwestern Missouri*. Report of Investigations, no. 50. Rolla: Missouri Geological Survey and Water Resources. 48 pp.

———. 1986. *Paleozoic Succession in Missouri, Part 4, Mississippian System*. Report of Investigations, no. 70. Rolla: Missouri Department of Natural Resources, Division of Geology and Land Survey. 182 pp.

Thompson, Thomas L., and L. D. Fellows. 1969. *Stratigraphy and Conodont Biostratigraphy of Kinderhookian and Osagean Rocks of Southwestern Missouri and Adjacent Areas*. Report of Investigations, no. 45. Rolla: Missouri Geological Survey and Water Resources. 263 pp.

Tolman, Carl, and Forbes Robertson. 1969. *Exposed Precambrian Rocks in Southeast Missouri*. Report of Investigations, no. 44. Rolla: Missouri Geological Survey and Water Resources. 68 pp.

Unklesbay, A. G. 1952. *The Geology of Boone County*. Rolla: Missouri Geological Survey and Water Resources. Vol. 33, 2d ser. 159 pp., 13 pls.

———. 1955. *The Common Fossils of Missouri*. Columbia: University of Missouri Press. 98 pp.

———. 1987. "Midwest Earthquakes." *Earth Science Magazine* 40, no. 4 (Winter): 11–13.

U.S. Senate. 1967. *Mineral and Water Resources of Missouri*. Senate Document No. 19., 90th Congress, 1st sess. 399 pp. Washington: U.S. Government Printing Office. Also published as Missouri Geological Survey, 2d ser., vol. 43, 1967.

Vineyard, Jerry D., and Gerald L. Feder. 1982. *Springs of Missouri.* Water Resources Report, no. 29. Rolla: Missouri Department of Natural Resources, Division of Geology and Land Survey. 212 pp.

Vineyard, Jerry D., and James H. Williams. 1965. "A New Sink in Laclede County, Missouri." *Missouri Mineral Industry News* 5, no. 7 (July): 69–71.

Weaver, H. Dwight, and Paul A. Johnson. 1980. *Missouri: The Cave State.* Jefferson City: Discovery Enterprises. 336 pp.

Weaver, H. Dwight, James N. Huckins, and Rickard L. Walk. *The Wilderness Underground: Caves of the Ozark Plateau.* Columbia: University of Missouri Press, 1992. 128 pp., 111 pls.

Wharton, Heyward M., et al. 1969. *Missouri Minerals—Resources, Production, and Forecasts.* Rolla: Missouri Geological Survey and Water Resources. Spec. pub. no. 1. 303 pp.

Whitfield, John W. 1981. *Underground Space Resources in Missouri.* Report of Investigations, no. 65. Rolla: Missouri Department of Natural Resources, Division of Geology and Land Survey. 65 pp., 49 figs.

The following field-trip guidebooks are published by the Missouri Division of Geology and Land Survey in Rolla and its predecessor agency, the Geological Survey. Those marked by an asterisk are out of print but can probably be found in college and university libraries.

*Guidebook for 16th regional field conference, Kansas Geological Society, West-Central Missouri, 1952. Report of Investigations, no. 13.

*Guidebook of 17th regional field conference, Kansas Geological Society, Southeastern and South Central Missouri, 1954. Report of Investigations, no. 17.

*Guidebook, second annual meeting, Association of Missouri Geologists, 1955. Report of Investigations, no. 20.

Guidebook to geology of St. Francois Mountains Area, 1961. Report of Investigations, no. 26.

Guidebook for 26th regional field conference, Kansas Geological Society, Northeastern Missouri and West-Central Illinois, 1961. Report of Investigations, no. 27.

*Guidebook for 1965 annual meeting of Geological Society of America. Cryptoexplosive structures in Missouri, 1965. Report of Investigations, no. 30.

Guidebook for 1965 annual meeting of Geological Society of America Kansas City Group at Kansas City, 1965. Report of Investigations, no. 31.

*Guidebook to Middle Ordovician and Mississippian strata in St. Louis and St. Charles Counties. 1966 annual meeting of American Association of Petroleum Geologists, 1965. Report of Investigations, no. 34.

*Guidebook to geology between Springfield and Branson, Missouri. Stratigraphy and cave development, 1976. Report of Investigations, no. 37.

Guidebook to geology and ore deposits of selected mines in the Viburnum Trend, Missouri, 1975. Report of Investigations, no. 58.

Guidebook to geology along Interstate 55 in Missouri, 1977. Report of Investigations, no. 62.

Guidebook to the geology and ore deposits of the St. Francois Mountains, Missouri, 1981. Report of Investigations, no. 67.

MUSEUM LIST

The following list indicates sites where many of the geological aspects of the state can be examined firsthand. Sites are arranged in alphabetical order by nearest city.

Name of Museum	Location	What to See
Geology Building	University of Missouri–Columbia	exhibits in lobby and hallways
Missouri Mines State Historic Site Geology Museum	St. Joe State Park, Flat River	museum of mining, geology, and mining history
Ralph Foster Museum	School of the Ozarks, Hollister	exhibits of rocks and minerals, plus items of historical interest
Mastodon State Park Museum and Visitor Center	Imperial	exhibits on Pleistocene geology and vertebrate paleontology; complete mastodon skeleton
Tri-State Mineral Museum	Joplin	mineral specimens
Geosciences Department Museum	University of Missouri–Kansas City	museum specializing in ammonite fossils
Kansas City Museum	Kansas City	exhibits on fossils
Onondaga Cave State Park Visitor Center	Leasburg	exhibits on caves and cave development, life, history, and environmental relationships
Bennett Spring State Park	12 miles west of Lebanon on Highway 64	exhibits on geology and hydrology of major spring systems

Name of Museum	Location	What to See
Geology Museum	McNutt Hall, University of Missouri–Rolla	mineral and rock exhibits
Ed Clark Museum of Missouri Geology	DGLS Buehler Building, Rolla	miscellaneous exhibits on Missouri geology, rocks and minerals, and paleontology; oldest geologic map of state
Maramec Spring Park	St. James	exhibits on iron mining, spring hydrology, general geology
Science Center	St. Louis	museum covering all phases of science; a major attraction for all ages
Department of Earth and Planetary Sciences	Washington University, St. Louis	exhibits in hallways, including rock, mineral, and fossil collections, and space science
Meramec State Park Visitor Center	Sullivan	exhibits showing relationships among geology, biology, history, prehistory, and paleontology; stratigraphy, caves, and rock types are emphasized; special emphasis on rivers, springs, geology, and hydrology
Taum Sauk Museum	Profitt Mountain on Taum Sauk Pumped Storage Power Plant property west of Ironton	general natural history exhibits; biological and geological; rocks and maps; geological cross section of Missouri
Truman Lake Visitor Center	Warsaw	mastodon and other Pleistocene vertebrate fossils

INDEX

187

ABOUT THE AUTHORS

A. G. Unklesbay is a retired Professor of Geology and a former Vice President of Administration for the University of Missouri System. He is a Fellow of the Geological Society of America and served as Executive Director of the American Geological Institute for six years. He is author of numerous books including *The Common Fossils of Missouri* (University of Missouri Press).

Jerry D. Vineyard is Deputy State Geologist of Missouri and Director of the Missouri State Water Plan. He has had a career-long fascination with the geology and hydrology of Missouri. He is author and coauthor of several books, including *Springs of Missouri* and *Geologic Wonders and Curiosities of Missouri*, Revised Edition.